Concepts of Tissue-Biomaterial Interactions

Concepts of Tissue-Biomaterial Interactions

Fundamentals and New Directions

Rena Bizios
Lutcher Brown Endowed Distinguished University Chair Professor
Department of Biomedical Engineering
The University of Texas at San Antonio
One UTSA Circle
San Antonio, TX, USA

Marissa E. Wechsler
Assistant Professor
Department of Biomedical Engineering and Chemical Engineering
The University of Texas at San Antonio
One UTSA Circle
San Antonio, TX, USA

WILEY

Published by John Wiley & Sons, Inc., Hoboken, New Jersey.
Published simultaneously in Canada.

***Library of Congress Cataloging-in-Publication Data applied for*:**

Hardback ISBN: 9781119841005

Cover Design: Wiley
Cover Image: © Tanvir Ibna Shafi/Getty Images

Set in 9.5/12.5pt STIXTwoText by Straive, Chennai, India

To all past, present, and future students and colleagues, our comrades (συνταξιδιώτες) in the exploration of the interactions of cells and tissues with implantable biomaterials and medical devices. Our combined, continuous interest and contributions have placed this field in the mainstream of biomedical engineering.

Contents

List of Figures

List of Tables

Preface

The original concept of *Tissue-Biomaterial Interactions* was developed into a new undergraduate course and was taught, for the first time, in 1989 at a time when biomedical engineering education was *very* young and needed everything, including new courses.

For that purpose, pioneering (for those times), interdisciplinary approaches judiciously chose and used fundamental concepts from physiology and biology and integrated them in a biomedical engineering course.

The aforementioned course was motivated by, and addressed, a need for pertinent textbooks in undergraduate biomedical engineering curricula. Judging by its subsequent appeal, acceptance, and use, the book *An Introduction to Tissue-Biomaterial Interactions* successfully met that need in academia.

In the interim, biomedical engineering has enjoyed remarkable growth and development. Thus, the current milieu needs an updated textbook which still addresses the process of wound healing and the latest developments regarding various interactions of the human body with implant biomaterials and medical devices. The present textbook is intended for a one-semester course during the third year of undergraduate curricula. By that time, students in a "typical biomedical engineering undergraduate curriculum" would have already taken introductory courses in biomaterials, cell biology, chemistry, and biomechanics; concepts of the aforementioned courses are the background needed to take a *Tissue-Biomaterial Interactions* course. In our opinion, a course using the present textbook is appropriate training for both undergraduate and first-year biomedical engineering graduate students focusing on biomaterials, cellular engineering, tissue engineering, and implantable medical devices.

This textbook summarizes our pertinent teaching experience with undergraduate and graduate courses to date. We are fully aware, however, of the future needs of biomedical engineering, a dynamic discipline which continuously expands into new scientific frontiers. For this reason, we consider this textbook only an introduction to a fascinating field of knowledge which provides fundamental training for students pertinent to their future careers. We encourage instructors, who use this textbook in teaching pertinent courses, to supplement the material presented in this textbook with appropriate updated information from the latest biomedical literature, as well as from their own research endeavors.

Acknowledgments

In writing this textbook, the result of teaching pertinent undergraduate and graduate courses, we benefited from the encouragement and support received from our students, colleagues, and families. We thank you all!

We wish, however, to especially acknowledge and thank Ms. Jordyn M. Wyse, for assistance with the figures of the textbook, and Ms. Leslie Sanchez, for assistance with various managerial aspects in finalizing this endeavor. All figures throughout this textbook were created with BioRender.com.

About the Companion Website

This book is accompanied by the companion website:

www.wiley.com/go/Concepts_of_Tissue-Biomaterial_Interaction_1e

The website includes:
Supplementary to script

1

Materials for Prosthetic Devices and Implants Through the Ages

Through the ages, human medical needs have resulted from birth defects, trauma, age-related diseases, degenerative conditions, end-stage organ failure, etc. Several biological species have the capability for tissue/organ regeneration; for example, salamanders can regenerate lost limbs, tails, and even eyes [1]; deer regenerate antlers [2]; flatworms regenerate their bodies [3]; rabbits, domestic cats, and bats fill in ear holes [4]; zebrafish regenerate fins, etc. [5]. In contrast, humans have an extremely limited capacity to regenerate body tissues and organs.

In this respect, the Greek myth of Prometheus is pertinent. Long time ago, only the Gods of Olympus were privileged to have the comfort of fire. During those times, the mortal humans were doomed to live in a dark and cold world. Prometheus, a Titan and immortal, commiserated with the humans, stole the fire from Olympus, and, in direct opposition to the will of Zeus, the father of the Olympian Gods, gave it (and, thus, hope) to humans. This defiant act enraged Zeus, who severely punished Prometheus to be helplessly chained on the mountain Caucasus while a portion of his liver was eaten by a bird of prey every day. Because the liver regenerated overnight, this situation provided endless food supply to the bird of prey but subjected Prometheus to eternal torture. This Greek myth makes a great story but is also notable because it is part of the tradition of a civilization, which emphasized exclusively thinking approaches versus experimentation. Most remarkable is the "kernel" of truth, which the Prometheus myth contains. Organ and tissue regeneration in humans are extremely limited with the following very few exceptions: liver is the only human organ, which can regenerate completely from a minimum of 25% of its original mass [6]. As far as other tissues are concerned, bone fractures (but not nonunion bone defects) and small skin cuts heal through respective tissue regeneration processes.

Concepts of Tissue-Biomaterial Interactions: Fundamentals and New Directions,
First Edition. Rena Bizios and Marissa E. Wechsler.

Companion Website: www.wiley.com/go/Concepts_of_Tissue-Biomaterial_Interaction_1e

1.1 Traditional External Prosthetic Devices

Through the millennia, medical responses to human loss of body parts (such as lower limbs and arms) were limited and involved exclusively external prostheses. Ancient treatments/solutions used "ordinary" materials such as wood, leather, string, etc., to assemble prostheses. An example is the replacement of the amputated right big toe of a 3000-year-old Egyptian mummy; this replaced appendage was composed of carved wood, attached to a piece of leather, and secured onto the foot of the recipient with string [7]. Undoubtedly, this is an anatomically correct and aesthetically acceptable prosthesis; its functional aspect, however, is questionable when one considers that the footwear of ancient Egyptians were sandals.

Prosthetic devices replacing lost parts of the body are also found in the literature. The legendary Captain Hook, in the children's fictional story of Peter Pan, was infamous for his bad deeds but also scary because he had a metal hook replacement for his lost left arm. Although anatomically incorrect, and aesthetically unacceptable, this prosthesis was most functional because it gave Captain Hook a reliable way to hold on (indeed for dear life) when his ship encountered bad weather in the open seas and, thus, prevented him from being swept away in the stormy waters.

As recently as the beginning of the 20th century, prosthetic limbs available to patients who needed them were composed of "ordinary" materials (such as wood and metals) and were anthropomorphic (with stiff and nonfunctional parts such as fingers) renditions of human arms and hands [8]. Synthetic materials (such as plastics and metals) that had been developed for various other industrial and engineering applications were used for making prosthetic devices. Attempts to incorporate functionality into prosthetic hands led to the inclusion of cylindrical tubes one each for the four fingers. Later on, incorporation of mechanical features enabled such prototypes of human hands to have motor control of the finger motions required for actions such as playing the piano [8]. Further advancements in design addressed the sensor needs of prosthetic hands. These devices used various materials and electronic parts and added a "thumb part", thus providing the hand prosthesis with the capability of grasp. During the 21st century, the needs of wounded soldiers promoted development of magnificent prosthetic arms and hands composed of "modern" materials (including various plastics, metals, and textiles); most importantly, these prostheses included sophisticated and advanced electronics attached to the muscles and neurons of the recipient amputee, thus allowing control of desirable movements and commands such as grasping, picking up objects, bringing a bottle to the mouth, and successfully drinking the contained water without any spills [9]. Another very important development of prosthetic hands was incorporation of mechanical and electronic parts that provided the recipient patients with the sensation of touch (i.e., recognizing hard and soft items) and pressure (e.g., while shaking hands) [10, 11].

During the aforementioned times, development of prosthetic legs paralleled that of arms and hands. Anthropomorphic but unyielding, one-piece (including monolithic wooden foot, calf, knee, and part of the thigh) legs were eventually replaced with devices which did not look like the human limbs but had unparalleled advantages. Such prostheses, composed of "modern" materials which are light and have the appropriate mechanical properties (such as strength in carrying mechanical loads during human locomotion, flexibility, etc.), provide needed support for several movements of the human body such as standing up and sitting down, walking (on various terrains including not only flat but also inclined surfaces and staircases) [12], jogging and running, swimming [13], and even surfing [14].

It should be noted that, in addition to humans, small (such as kittens) and big (such as elephants) animals need, and thus have received, prosthetic devices for missing or injured limbs [15]. Such prostheses are appropriately designed and scaled, either up or down, and are composed of similar materials as those used for applications to human patients. Veterinary care of pets, animals used in racing events, animals kept in zoos, and farm animals constitutes a sizable and major market for veterinary care and many biomedical applications including prosthetic devices.

1.2 Implantable Devices

External prosthetic devices have helped patients, but they neither addressed nor resolved medical needs requiring replacement, healing, and regeneration of mammalian tissues and organs; such needs require implantable devices. Both external and implantable prostheses are composed of materials and contain combinations of mechanical and electrical parts. In contrast to the temporary and intermittent use of the traditional, external prosthetic devices, implants are placed inside the bodies of patients (humans and/or animals) for various but mostly long-term time-periods, specifically the lifespan of the recipient.

Initially, synthetic materials developed for non-biomedical applications were chosen and used for biomedical implants; pertinent examples include elastic undergarment textiles to produce synthetic tubular substitutes for blood vessels, and titanium/titanium alloys (relatively light but strong materials) for load-carrying orthopedic implants (such as total hip prostheses).

Because such cavalier, *trial-and-error* approaches resulted in dramatic failures (including implant malfunction and rejection), it became obvious that successful biomedical implant materials and devices require new specifications and must fulfill different criteria. Implant biomaterials (the components of implant devices) were eventually designed to be nontoxic, non-pyrogenic, nonallergenic, etc. Adherence to, and applications of, these requirements resulted in avoidance of implant rejection. For a while, attainment of a state of "tolerance" by the surrounding milieu inside the recipient patients' bodies became an acceptable clinical outcome, despite the fact that those implants did not produce desirable and controlled interactions with surrounding tissues and did not trigger the normal tissue-healing process at the implant–tissue interface.

1.3 Materials Used for Implants

By definition, a "biomaterial" is either a natural or synthetic material used for biomedical applications to support, enhance, or replace damaged tissue or a biological function [16].

A summary of the types of materials which have been used for implants is given in Table 1.1.

The historical development and progression of biomaterials were appropriately summarized by Dr. James A. Anderson, a pioneer, world-renowned pathologist, and biomaterials expert, who stated that the relative focus

Table 1.1 Types of materials used for implants.

Implant material	
Polymers	• Natural (e.g., collagen, fibrin, etc.) • Products of organic chemistry synthesis • *Special formulation*: Biodegradable polymers
Ceramics	• Natural (e.g., devitalized bone, coral, etc.) • Products of inorganic chemistry synthesis
Metals	• Ores
Composites	• Products of two or more the above materials • Products of various chemistry methods

Important Requirement: Approval by the Food and Drug Administration (FDA) in the United States of America and by pertinent regulatory agencies in other countries around the world is required for implant materials used in biomedical applications.

on the "bio" and "materials" components of the term "biomaterials" has changed over the years and can be classified into three historical eras of this field. Specifically, the materials aspect was the focus of pertinent research and development before 1975, followed by a period when both the "bio" and "materials" aspects attracted and received pertinent attention from researchers. Since 2000, however, the predominant focus of the biomaterials field has been on the "bio" component [17].

The most desirable outcome, and simultaneously the major biomedical need and challenge pertinent to implant materials and devices, is formation of normal and functional new tissue, eventually leading to timely and uncomplicated integration of synthetic implants into surrounding tissues in vivo. Such approaches depend on, and benefit from, continuous advances in several scientific and engineering fields including physiology, medicine, cellular and molecular biology, genetics, extracellular matrix biology, biochemistry, biomaterials science and engineering, cellular and tissue engineering, biomedical engineering, pathology, and regenerative medicine.

Presently, application of accumulated research-related experience is focusing on incorporating the latest knowledge from advances in several scientific and engineering fields regarding the design, formulation, and extensive testing of implantable materials and devices for specific clinical applications to meet biocompatibility requirements, leading to their approval by pertinent regulatory agencies (such as the Food and Drug Administration (FDA) in the United States and other pertinent regulatory agencies in other countries around the world) [18]. The process by which a material becomes a "biomaterial" for implants is continuously evolving and improving as it encompasses the latest developments from several pertinent scientific and engineering fields.

References

1 Joven A, Elewa A, Simon A. Model systems for regeneration: salamanders. *Development* [Internet]. 2019 Jul 1 [cited 2022 Jan 24];146(14). Available from: https://www.ncbi.nlm.nih.gov/pmc/articles/PMC6679358/.

2 Li C, Zhao H, Liu Z, McMahon C. Deer antler—a novel model for studying organ regeneration in mammals. *Int. J. Biochem. Cell Biol.* [Internet]. 2014;56:111–22. Available from: https://pubmed.ncbi.nlm.nih.gov/25046387/.

3 Tanaka EM, Reddien PW. The cellular basis for animal regeneration. *Dev. Cell* 2011 Jul 19;21(1):172–85.

4 Metcalfe AD, Willis H, Beare A, Ferguson MWJ. Characterizing regeneration in the mammalian external ear. *J. Anat.*2006 Oct [cited 2022 Jan 24];209(4):439. Available from: https://www.ncbi.nlm.nih.gov/pmc/articles/PMC2100363/.

5 Pfefferli C, Jaźwińska A. The art of fin regeneration in zebrafish. *Regeneration* 2015 Apr [cited 2022 Jan 24];2(2):72. Available from: /pmc/articles/PMC4895310/.

6 Sheedfar F, Di Biase S, Koonen D, Vinciguerra M. Liver diseases and aging: friends or foes? *Aging Cell* [Internet]. 2013 Dec [cited 2022 Jan 24];12(6):950–4. https://pubmed.ncbi.nlm.nih.gov/23815295/.

7 This 3,000-Year-Old Wooden Toe Shows Early Artistry of Prosthetics | Smart News | Smithsonian Magazine, 2017. [Internet]. [cited 2022 Jan 24]. Available from: https://www.smithsonianmag.com/smart-news/study-reveals-secrets-ancient-cairo-toe-180963783/.

8 James R, Laurencin CT. Regenerative engineering and bionic limbs. *Rare Metals.* 2015 34(3):143–155. https://www.ncbi.nlm.nih.gov/pmc/articles/PMC4429301/.

9 Vox. FDA Approves Robotic Prosthesis Controlled by Muscle Contractions – Vox [Internet]. [cited 2022 Jan 24]. Available from: https://www.vox.com/2014/5/11/11626740/fda-approves-robotic-prosthesis-controlled-by-muscle-contractions.

10 Inhabitat. Shaking hands with the LifeHand 2 | Inhabitat – Green Design, Innovation, Architecture, Green Building [Internet]. [cited 2022 Jan 24]. Available from: https://inhabitat.com/thought-controlled-robotic-arm-returns-the-sense-of-touch-to-amputees/lifehand2-1/.

11 Raspopovic S, Capogrosso M, Petrini FM, Bonizzato M, Rigosa J, Di Pino G, et al. Bioengineering: restoring natural sensory feedback in real-time bidirectional hand prostheses. *Sci. Transl. Med.* [Internet]. 2014 [cited 2022 Jan 24];6(222). Available from: https://www.newscientist.com/article/dn25008-natural-sense-of-touch-restored-with-bionic-hand/.

12 BBC (2014). Lesser-known things about prosthetic legs – BBC News [Internet]. [cited 2022 Oct 31]. Available from: https://www.bbc.com/news/blogs-ouch-28225622.

13 Goldstein T, Oreste A, Hutnick G, Chory A, Chehata V, Seldin J, et al. A pilot study testing a novel 3D printed amphibious lower limb prosthesis in a recreational pool setting. *PM&R* [Internet]. 2020 [cited 2022 Jan 24];12(8):783. Available from: /pmc/articles/PMC7496828/.

14 The Nexus (2021). Designing a Prosthetic to Help Surfers Catch a Wave – The Nexus [Internet]. [cited 2022 Jan 24]. Available from: https://nexus.jefferson.edu/design-and-style/designing-a-prosthetic-to-help-surfers-catch-a-wave/.

15 ABC News (2014). Double-Amputee Cat "Paw Stands" Down Stairs – ABC News [Internet]. [cited 2022 Oct 31]. Available from: https://abcnews.go.com/Lifestyle/Pets/double-amputee-cat-paw-stands-stairs/story?id=23991141.

16 NIBIB (2017). Biomaterials [Internet]. [cited 2022 Jan 24]. Available from: https://www.nibib.nih.gov/science-education/science-topics/biomaterials.

17 Anderson JM. The future of biomedical materials. *J. Mater. Sci. Mater. Med.* 2006;17:1025–8.

18 Williams DF. The biomaterials conundrum in tissue engineering. *Tissue Eng. Part A* 2014;20:1129–31.

2

Lessons Learned from Physiology

2.1 Tissue Wound Healing: Introduction

Implantation of biomaterial-containing prosthetic medical devices into the mammalian body (a well-vascularized and innervated entity) requires surgical procedures with unavoidable outcomes: injury to several tissues (such as skin, fat, muscle, blood vessels, nerves, and other tissues pertinent to specific anatomical locations), which are cut by the surgeon's scalpel. Such injury affects detrimentally all tissue constituents and involves (1) cell death, (2) denaturation/destruction of the extracellular matrix, (3) loss of blood vessel integrity, and (4) damage to nerves. Subsequent, successful tissue wound healing requires cell regeneration, synthesis, and deposition of components of a new extracellular matrix, formation of new blood vessels, and restoration of nerves (Figure 2.1).

Physiology provides valuable information regarding the tissue wound healing process. Healing of injured tissues in humans is limited and has diverse outcomes. Skin, whose healing has been studied extensively and is the model tissue in this field, can heal small (on the order of several millimeters) cuts. In such cases, the healing process involves "bridging" over the injury site, a process facilitated by bringing together "similar" tissue components (i.e., cells, extracellular matrix, biological liquids, etc.) from the perimeter of the injury (Figure 2.2). In contrast, and in addition to the unavoidable induced trauma/injury, surgical introduction of implants into the body creates a tissue/implant interface; the presence of a biomaterial on one side of the gaping injury site deters the "normal" tissue wound healing process. A comparison of the details on the two sides of the tissue/implant interface highlights the main difference: there is a biological extracellular matrix on one side but an implant biomaterial on the other (Figure 2.3). The wound healing process and outcome under these

Concepts of Tissue-Biomaterial Interactions: Fundamentals and New Directions,
First Edition. Rena Bizios and Marissa E. Wechsler.

Companion Website: www.wiley.com/go/Concepts_of_Tissue-Biomaterial_Interaction_1e

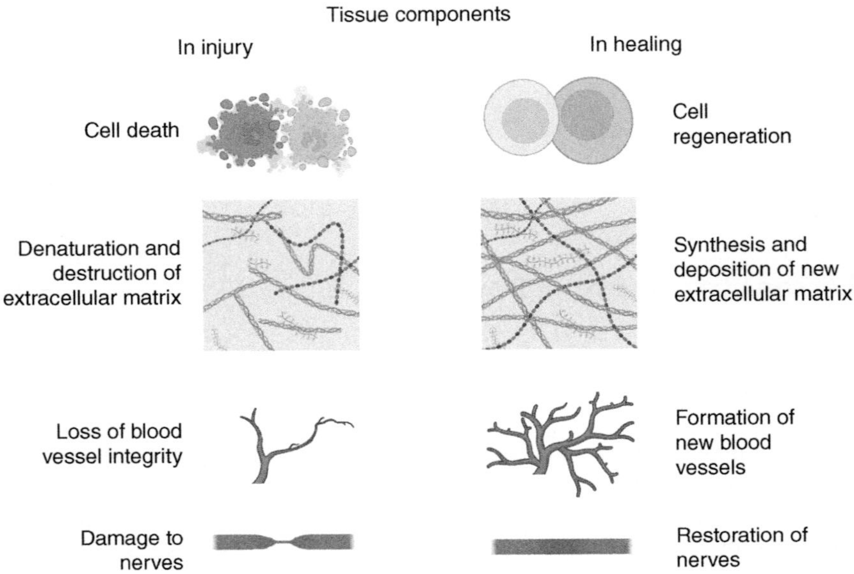

Figure 2.1 Schematic outline of tissue components affected by injury (left side) and repaired during subsequent wound healing (right side).

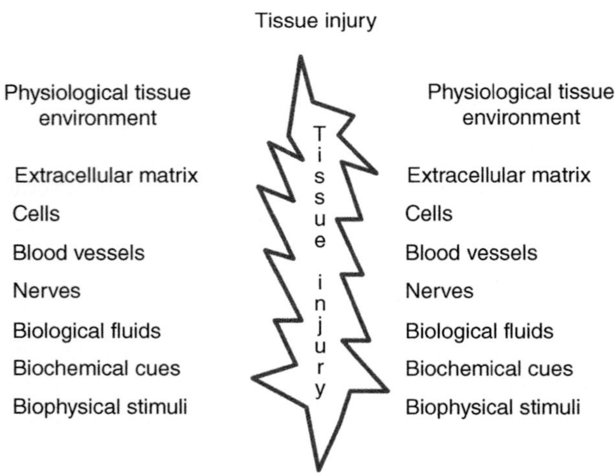

Figure 2.2 The physiological tissue environment in the case of normal tissue injury.

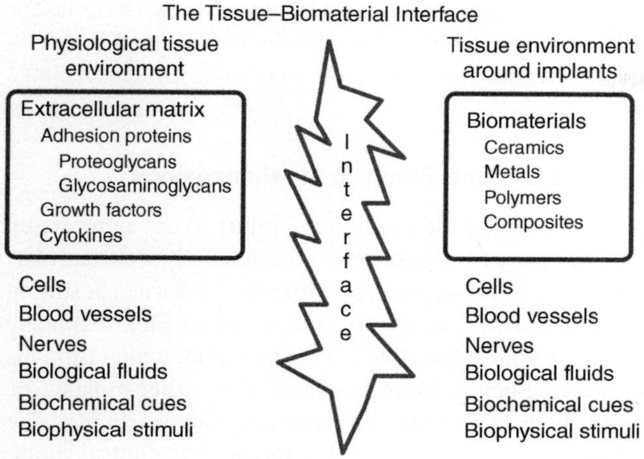

Figure 2.3 Comparison of the major tissue components at the interface between injured tissues and implants.

circumstances control the fate of implants and are, thus, of paramount importance to surgeons, physicians, biomedical engineers, and implant biomaterial scientists/engineers.

2.2 The Process of Normal Wound Healing in Injured Tissues

Following tissue injury, the normal wound healing process proceeds via several stages, which are partially overlapping, parallel, and synchronous but mostly sequential. These stages are the following: Blood Coagulation and Fibrinolysis, Inflammation, Proliferation/Repair, and Remodeling/Maturation (listed in Table 2.1). Under normal circumstances, the duration

Table 2.1 Stages of the normal wound healing process following tissue injury.

Blood coagulation and fibrinolysis
Inflammation
Proliferation/repair
Remodeling/maturation

of each stage of the wound healing process varies from up to two minutes for Coagulation, from few hours to few days for Inflammation, from days to two weeks for Proliferation/Repair, and from weeks to months for Remodeling/Maturation.

2.2.1 The Importance of Blood Supply in the Human Body

Even when not directly aimed at the vasculature, injury to any other tissue and organ in the human body (exceptionally and extensively well vascularized) involves damage to blood vessels (predominantly capillaries), resulting in blood pouring outside of the circulatory system. Blood (approximately 5 liters in adults) [1–3] is a most valuable biological liquid, which humans need for survival and well-being. Excessive blood loss, either outside the body or to parts of the body other than the cardiovascular system (e.g., the brain and the peritoneal cavity), can lead to life-threatening clinical conditions and have serious health-related consequences.

Blood supply is of crucial importance in the human body because cells require oxygen and nutrients to survive and function. The aforementioned life-supporting chemicals are provided by the circulating blood, which also removes carbon dioxide and metabolic waste from the cells, tissues, and organs. In order to survive, cells in the body must be located within a 100–200 μm distance from blood vessels and, thus, from the precious blood supply [4]. This requirement is successfully met in the well-vascularized tissues and organs in the human body.

2.2.2 Blood Composition

Blood is a biological liquid, which is also classified as a "tissue," a very special one indeed, because it contains cells suspended in a liquid with dissolved chemical compounds, such as proteins. Circulating human blood contains approximately 45% cells (packed volume) and 55% of a liquid called "plasma." Plasma, in turn, is composed of approximately 93% water and 7% dissolved chemicals, specifically proteins (of various molecular weights, sizes, and structures), ions, etc. Albumin is the most abundant, about 3.5% of the protein content in plasma, which also contains coagulation proteins (fibrinogen and coagulation factors), as well as globulins, complement proteins, and lipoproteins (Table 2.2). Each plasma protein has a distinct function: albumin is an important contributor to the body's osmotic equilibrium; fibrinogen and the coagulation factors are major contributors to blood coagulation, while activation of the complement proteins and globulins affects white blood cell functions pertinent to the body's defense against disease-causing pathogens [1–3].

Table 2.2 Composition of human blood plasma.

Blood plasma components	Weight percent
Water	92
Proteins[a)]	7
Albumin	58
Globulin	37
Fibrinogen	4
Other proteins	1
Other solutes (such as electrolytes, nutrients, cell metabolic waste, respiratory gases)	1

a) 7% total protein content in plasma.

Circulating blood contains three different types of cells: red blood cells or erythrocytes (characteristic dimension 6–8 μm), platelets or thrombocytes (characteristic dimension approximately 2–3 μm), and white blood cells or leukocytes (characteristic dimension 14–16 μm) [1]. In contrast to the one and only type of red blood cells and of platelets, the circulating white blood cell "family" contains the following cell types: neutrophils (or polymorphonuclear leukocytes), lymphocytes, monocytes, eosinophils, and basophils (Table 2.3). Each blood cell type has a predominant and specific function: red blood cells carry oxygen to, and remove carbon dioxide from, all parts of the body; platelets are participants and major contributors to the blood coagulation (thrombosis) process; and white blood

Table 2.3 Cells in human whole blood [1–3].

Cells	Average cell number (per mm^3)
Erythrocytes (red blood cells)	4,200,000–6,200,000
Thrombocytes (platelets)	120,000–300,000
Leukocytes (white blood cells)	5,000–10,000
Neutrophils	40–70% (of total leukocytes)
Lymphocytes	20–40% (of total leukocytes)
Monocytes	2–8% (of total leukocytes)
Eosinophils	1–4% (of total leukocytes)
Basophils	0.5–1% (of total leukocytes)

cells (the defenders of the human body against invading, disease-causing pathogens) are also major contributors to the Inflammation and participants in the Fibrinolysis stages of the tissue wound healing process [1–3].

2.2.3 Hematopoiesis (Formation of Blood Cells)

The bone marrow, the site of all blood cell formation, is rich in progenitor and stem cells, which are located in hematopoietic regions enclosed within porous membranes. During the embryonic stage, the marrow of all bones is the place of blood cell production. After birth, and especially in adulthood, blood cells are produced only in the red marrow located in the flat bones (specifically the calvarium, sternum, ribs, and the epiphyses of long bones) [5]. Of the processes which result in new blood cell formation, the mechanism of red blood cell formation was the first to be elucidated [6]. Detection of low oxygen content in blood, as it is filtered in the kidneys, instigates release of "erythropoietin", a chemical compound which is transported to the bone marrow and commits erythrocyte progenitors to new red blood cell formation through a multistep process. During release into the circulation, the red blood cells lose their nucleus, which, because of its large size, cannot pass through the pores of the membrane which encloses the hematopoietic region and is, thus, left behind inside the marrow. Immature red blood cells, known as "reticulocytes" (identifiable by their ribosomal RNA content), may also be released, but they mature within a day in the circulation.

Similar mechanisms involving other pertinent chemical compounds are responsible for starting the process of platelet and white blood cell formation [7]. Unique cells in this process of platelet formation are the *megakaryocytes*, giant cells with big, lobulated nuclei, present in the marrow during the final stage of platelet formation. Finger-like extensions of megakaryocyte cytoplasm penetrate the pores of the membrane surrounding the hematopoietic regions; their ends are small, anucleate blebs that are pinched off and released into the blood circulation as new platelets [7].

Formation of white blood cells (leukocytes) from committed multipotent progenitor cells in the bone marrow involves several differentiation processes controlled and modulated by various chemical compounds (such as select interleukins). Formation of neutrophils exemplifies this stepwise process: granulocyte–monocyte progenitor cells differentiate into promyelocytes, myelocytes, metamyelocytes, and, finally, into band neutrophils; during this process, neutrophils acquire their characteristic nuclear lobular morphology and various granules. The granules in the cytoplasm contain membrane-bound enzymes (e.g., proteolytic enzymes, lysozyme, cathepsin

G, myeloperoxidase, etc.) [8]. Leukocytes are the only blood cells with nuclei. In fact, the nucleus morphology (e.g., circular, lobulated, etc.) and nucleus size (relative to the cytoplasm and respective cell dimensions), as well as the presence of granules, are markers used to identify and classify the various white blood cells circulating in the human blood. Circulating human blood contains approximately 40–60% neutrophils, 20–40% lymphocytes, 2–8% monocytes, 1–4% eosinophils, and 0–1% basophils (Table 2.3) [2, 3].

With the exception of red blood cells, which are "red" due to their hemoglobin content, platelets and all types of white blood cells are colorless. In laboratory preparations, platelets and white blood cells are treated with various judiciously chosen stains prior to visualization using microscopy techniques. These methods allow identification of the various types of white blood cells (e.g., neutrophils versus monocytes, etc.) based on the nucleus morphology (e.g., circular, lobulated, etc.) and nucleus size (relative to the cytoplasm and respective cell dimensions), as well as the presence of granules; these details are markers used to identify and classify the various white blood cells circulating in the human blood. For example, under pertinent preparation and visualization, neutrophils are identified by their lobulated nucleus and granules dispersed in their cytoplasm; monocytes have a "bean-shaped" nucleus and lightly stained granulocytes; eosinophils have many rose-stained granulocytes, basophils have many dark blue-/purple-stained granulocytes, and lymphocytes have big (with respect to the overall cell dimension), round nuclei surrounded by a thin band of cytoplasm (Figure 2.4).

Under normal physiological conditions in humans, the newly released red blood cells (round with a characteristic "dimple" in the middle) circulate for up to 120 days; the platelets (discoid in shape) circulate for up to 10 days, and the white blood cells (spherical of various sizes depending on the specific subgroups) circulate for hours up to days. At the end of their respective lifespans, old and damaged blood cells are removed from the blood circulation and eliminated by phagocytes in the spleen and liver;

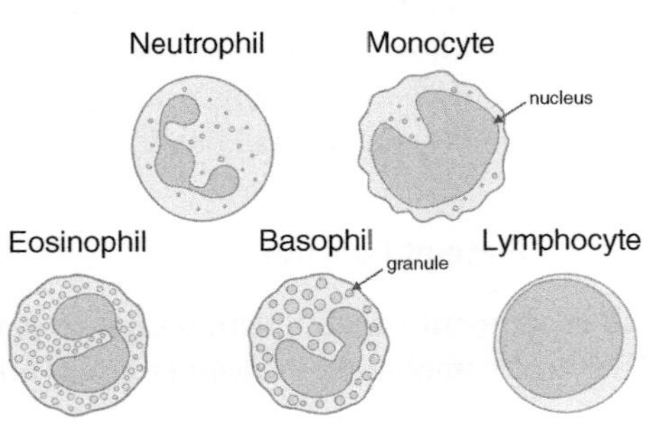

Figure 2.4 Schematic illustrations (not to scale) of leukocyte morphology and content.

during this process, some contents (e.g., iron from the red blood cells) are recycled. Participation in the functions (for which each blood cell type was created) as well as pathological conditions shorten the aforementioned life span of the circulating blood cells.

2.3 Blood Clotting (Thrombosis)

Blood is a unique liquid. In fact, blood is a non-Newtonian fluid because it is a suspension of various formed elements (i.e., blood cells) in plasma (mostly water containing several dissolved proteins, electrolytes, etc.).

Under normal physiological conditions, blood behaves as a "liquid" and flows within the circulation. Under special circumstances, however, this "liquid" becomes a "special" solid known as "blood clot" (thrombus). Among the circulating blood cells, platelets are the ones which actively participate in, and contribute to, the process of blood clotting or thrombosis.

2.3.1 Platelet Anatomy

Throughout their lifespan, approximately 75% of the released platelets are found in the blood circulation and 25% reside in the spleen [9]. Under normal physiological conditions, these small (characteristic dimension of approximately 2–3 μm), discus-shaped, anucleate cells circulate not activated. The latter condition is manifested by their monodispersed presence in circulating blood, the smooth, exterior plasma-membrane perimeter, by their microtubules (a set of concentric, small-diameter tubes in the cytoplasm located close to the cell perimeter), by special, compartments inside their cytoplasm known as "alpha granules" (containing adhesive proteins, clotting factors, growth factors, chemokines, cytokines, and glycosaminoglycans), and by "dense bodies" (containing serotonin, histamine, adenosine diphosphate (ADP), adenosine triphosphate (ATP) and calcium ions). Another unique platelet characteristic is several plasma membrane extensions known as the "surface-connected open canalicular system" composed of channels, which are open to the external cell environment and extend into the interior but do not perforate the platelets.

2.3.2 Platelet Function

An experiment which illustrates the response of platelets to select soluble chemical compounds (*activating agents*) *in vivo* can be conducted in the laboratory (*in vitro*) using a suspension of monodispersed platelets in a buffer. Chemical compounds such as epinephrine, serotonin, ADP, thrombin, and soluble collagen are known to be released from injured tissues, during several pathological conditions, from activated platelets, and during the plasma coagulation process. Addition of ADP, for example, to a suspension of platelets in buffer induces an immediate response (starting within a few seconds from the moment of exposure to the added chemical compound) and is manifested by profound morphological changes of the responding platelets. The perimeter of the plasma membrane develops several finger-like cytoplasmic projections (*filopodia*) with which platelets attach to other activated platelets and coalesce (*platelet aggregation*). In fact, the original monodispersed platelet suspension becomes an amorphous, condensed mesh (similar to a small clump of "glue" in the water part of the buffer) within 60–90 seconds. Subsequent action of the actin–myosin content of the platelets causes contraction of this mesh. The aforementioned external morphological changes involve simultaneous changes in the platelet interior: the microtubules contract toward the cell center, bringing the chemical-containing granules next to the surface-connecting open canalicular system, squeezing the contents of the granules into the canalicular channels, and thus discharging them into the immediate environment of the activated platelets (*platelet release*). The soluble chemical compounds thus released from a few activated platelets trigger activation of many more platelets.

In addition to release of the chemical compounds stored in the platelet granules during aggregation, there is *de novo* synthesis of certain bioactive chemical compounds of the arachidonate family. Platelets have the enzymes to convert arachidonic acid released from the membranes of aggregated platelets into several chemical compounds; the major product in this respect is Thromboxane A_2, the most potent naturally occurring platelet-aggregating agent and, at the same time, a vasoconstrictor.

The platelet response to activating agents is irreversible: in fact, it is the end of the life of individual platelets but manifests their ultimate contribution to a very important physiological process: blood coagulation (thrombosis).

2.3.3 Platelet Function in Hemostasis

When platelets in circulating blood *in vivo* are exposed to the bioactive chemical compounds released from damaged vascular tissue (specifically, injured or missing endothelium), they become activated. Activated platelets use the

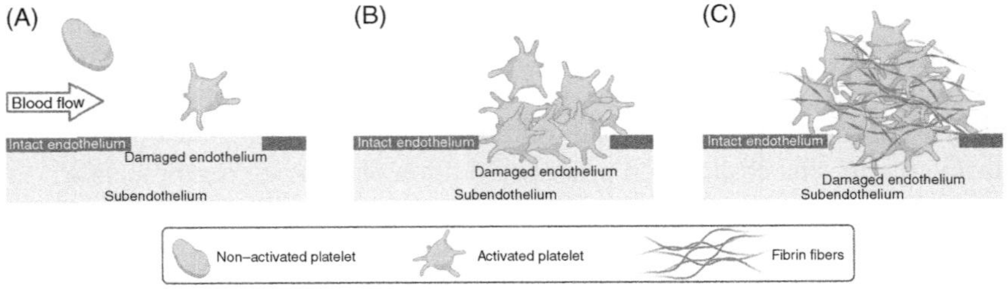

Figure 2.5 Schematic illustration (not to scale) of platelet plug formation. (A) Damage to endothelium activates platelets, evidenced by changes in their morphology. (B) Adhesion at the damaged endothelium site is followed by aggregation and formation of the platelet plug. (C) The fibrin network (end result of the plasma coagulation cascade) stabilizes the platelet plug and ensures the clot remains at the site of each formation in the dynamic environment of flowing blood.

ensuing filopodia to adhere to the injured tissue site (*platelet adhesion*), as well as to other activated platelets (*platelet aggregation*), release chemicals from their granules (*platelet release*), activate more platelets (via the released bioactive chemical compounds from their granules as well as by the newly synthesized and released Thromboxane A_2), pile up, and cover the injured tissue site (Figures 2.5A,B). In this case, the consolidated, gel-like "blob" of the aggregated platelets is known as "the platelet plug". In physiology, the platelet plug has two important purposes: (i) to prevent blood loss both outside the circulation (in parts of the human body, such as the peritoneal cavity, where blood should *not* be present) as well as outside the body; and (ii) to give the underlying injured tissues the time needed to heal.

Formation of the platelet plug is the ultimate purpose of the existence, and the crucial contribution of platelets, to blood homeostasis. In the dynamic environment of flowing blood in the vasculature, however, the platelet plug is unstable; it can be either chipped off or completely dislodged and then can travel downstream within the cardiovascular system. Ultimately, stability of the platelet plug is provided by the fibrin net, which is the outcome of the concurrent activation and process of the Plasma Coagulation Cascade (Figure 2.5C).

2.3.4 The Plasma Coagulation Cascade

Under normal physiological conditions, the coagulation proteins and coagulation factors are not-activated chemical compounds "dissolved" in the circulating blood plasma. The nomenclature used in hematology, as well as the synonyms by which these chemical compounds are known, is given in Table 2.4.

Table 2.4 Coagulation factors in human blood plasma.

Symbol	Synonym
I	Fibrinogen
II	Prothrombin
III	Tissue factor/tissue thromboplastin
IV	Calcium
V	Proaccelerin/labile factor
VII	Proconvertin/stable factor
VIII	Antihemophilic factor A
IX	Christmas factor
X	Stuart Prower factor
XI	Plasma thromboplastin antecedent
XII	Hageman factor
XIII	Fibrin stabilizing factor

Contact of circulating blood with injured tissues triggers a cascade of chemical reactions, which involve the coagulation proteins and coagulation factors outlined in Figure 2.6. The plasma coagulation cascade involves two pathways. Release of chemical compounds from the injured tissues (collectively known as "tissue thromboplastin" or "tissue factor"), in the presence of calcium ions, initiates the *Extrinsic Pathway* by activation of Factor VII (indicated as $FVII_a$), which in turn activates FX_a (the central step in the plasma coagulation cascade).

In contrast, the *Intrinsic Pathway* involves several steps before activation of FX. This pathway is initiated by activation of Factor XII (also known as the "Hageman Factor" or "Surface Factor"). Activation to Factor XII_a is achieved by several chemical compounds (such as collagen, ADP, kaolin, high molecular weight kininogen, prekallikrein, and other chemicals released by activated platelets). In addition, FXII is activated when, in the presence of collagen (and with contributions of bioactive chemical compounds released from activated platelets), blood comes into contact with the surfaces of materials *other than* the normal, intact, healthy endothelium (the innermost lining of the cardiovascular system and the *only natural, completely blood-compatible substrate*). In the presence of calcium ions, $FXII_a$ activates FXI. In the presence of calcium ions, FXI_a activates FIX; subsequently, FIX_a, along with $FVIII_a$, calcium ions, and phospholipids, activates FX [10].

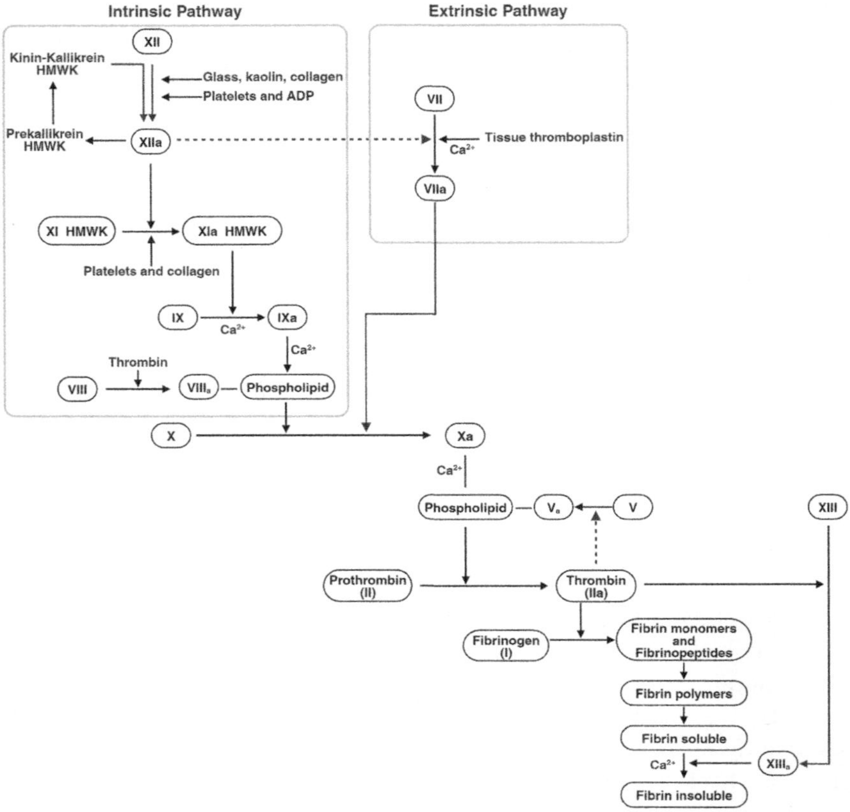

Figure 2.6 Flow chart of the plasma coagulation cascade. ADP, adenosine diphosphate; HMWK, high molecular weight kininogen; Ca^{2+}, calcium.

Prothrombinase, along with calcium ions, phospholipids (e.g., from platelets), FV_a, and FX_a, catalyzes the conversion of prothrombin (a precursor, inactive chemical compound in blood plasma) to thrombin. Thrombin catalyzes conversion of fibrinogen into fibrin. At first, the newly formed fibrin monomers are soluble, but in the presence of calcium ions and $FXIII_a$ (the thrombin-activated form of FXIII), fibrin becomes insoluble and forms a "net" dispersed throughout, and encompassing the platelet plug. This outcome stabilizes the platelet plug to withstand the mechanical forces in the dynamic milieu of the circulating blood. Inevitably, red blood cells "innocent bystanders" are trapped in large numbers inside these fibrin nets. The combined outcomes of the processes of the platelet plug and formation of the fibrin net result in the production of blood clots (thrombi).

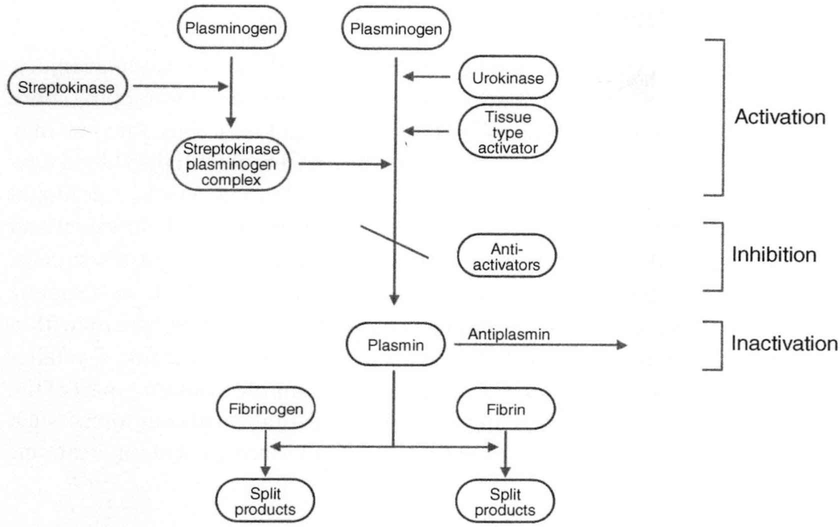

Figure 2.7 Flow chart of the conversion of plasminogen to plasmin.

2.3.5 "Synergisms" in the Process of Thrombus Formation

A notable aspect of the processes of platelet plug formation and the plasma coagulation cascade is several synergisms. Thrombin, an enzyme that catalyzes the conversion of inactive fibrinogen into fibrin (the grand finale of the plasma coagulation cascade), is also a potent platelet-aggregating agent and inducer of the platelet release reaction. Platelets contribute to various stages of the intrinsic pathway of the plasma coagulation cascade, specifically activation of coagulation factors FXI and FX, as well as a component of the prothrombinase complex (Figure 2.6).

In the broader physiological context, the process of thrombus formation initiated when blood comes in contact with what it considers "foreign surfaces" (i.e., anything other than normal, intact, healthy endothelium) is not an isolated event. In fact, activation of the coagulation Factor XII (the initiation step of the Intrinsic Pathway of the plasma coagulation process) instigates several other important processes. The process of fibrinolysis is initiated (Figure 2.7); the kallikrein/kinin (Figure 2.6) system (a contributor to the physiological processes associated with the sensation of pain) is activated; the complement proteins in the blood plasma, which trigger activation of white blood cell functions pertinent to inflammation, are also activated.

2.3.6 The Fate of Thrombi

Once a stable thrombus (blood clot) is formed at the site of vascular injury, it may grow (due to contributions by more platelets and by trapping more red blood cells as they pass by in the circulating blood). This outcome may cause clinical complications if the size of the growing clot hinders blood flow by narrowing (stenosis) the internal diameter of blood vessels. Additional thrombus growth, resulting in complete occlusion of the blood vessel and stopping blood flow, is detrimental because it deprives the downstream cells, tissues, and organs of their blood supply. Alternatively, in the hemodynamic milieu, a piece of a thrombus may be chipped off, carried downstream within the blood circulation, and may subsequently lodge itself inside a smaller diameter blood vessel, and thus, stop the blood supply to another part of the body. Such thrombotic developments lead to pathological conditions (such as heart attacks and strokes) and thus have important clinical implications.

2.3.7 Control of Thrombosis

A remarkable example of the "checks-and-balances" approach prevalent in human physiology is the natural control of the process of thrombosis. Without appropriate natural controls, once a thrombus is initiated, it may continue growing forever. The pertinent control process involves antithrombin III and heparin (present in human blood plasma), two chemical compounds dissolved in the human blood plasma. Antithrombin III inactivates thrombin in a slow chemical reaction. When, however, antithrombin III forms a complex with heparin, inactivation of thrombin proceeds very fast. Moreover, the antithrombin III-heparin complex inactivates all activated coagulation factors, specifically $FVII_a$, $FXII_a$, FXI_a, FIX_a, and FX_a, effectively cancelling all steps of the plasma coagulation cascade which precede formation of thrombin (Figure 2.6).

2.4 Fibrinolysis

Another physiological control of unwanted thrombi/blood clots is their removal. In physiology, once formed, thrombi remain at the sites of vascular injury to prevent blood loss from the circulation and to give the underlying injured tissue time to heal. Subsequently, each thrombus disintegrates and disappears via the physiological process of Fibrinolysis.

Fibrinolysis involves the enzymatic breakdown of the fibrin net, the stabilizing component of thrombi. This process is initiated by the conversion of plasminogen (a chemical compound dissolved in circulating blood plasma in an inactive form) to plasmin, a bioactive enzyme that degrades fibrin (Figure 2.7). This chemical reaction is catalyzed by the action of naturally produced enzymes such as urokinase (removed from blood while it is filtered in the kidneys and then excreted in urine) and tissue plasminogen activator (a bioactive compound produced by activated vascular endothelial cells). Once produced, plasmin breaks fibrin (in the blood clot) into successively smaller molecular weight fragments (known as "fibrin degradation products" or "fibrin split products"). Specifically fibrinogen (molecular weight of 340,000) is broken into fragment X (molecular weight of 240,000), then into fragments Y and D (molecular weights of 150,000 and 80,000, respectively), and finally into fragments E and D (molecular weights of 53,000 and 80,000, respectively) [11]. The small molecular weight fragments are removed from the body via natural excretion processes.

The fibrinolytic process is controlled by the inhibitory action of antiplasmin, a chemical compound dissolved in the blood plasma, which inactivates plasmin and, thus, stops unnecessary perpetuation of the fibrinolytic process (Figure 2.7).

The important outcome of fibrinolysis is removal of thrombi (blood clots) once the purpose of their formation and existence in the human body is successfully completed.

2.5 Inflammation

2.5.1 Introduction

Inflammation is a complex physiological response of vascularized tissues in the human body to injury and infection. The process involves activation of the immune system (including leukocytes, specific bioactive chemical compounds, and various proteins such as antibodies) to fight against, and thus protect, the human body from "foreign invaders" such as disease-causing pathogens (e.g., bacteria, viruses, and fungi), harmful "foreign intruders" (including splinters, thorns, etc.), as well as injuries following human body exposure to harmful chemical and physical conditions. In order to successfully defend the human body, the inflammation process attacks and eliminates the causes of tissue injury, clears out dead tissue (specifically cells and components of their extracellular matrix), and triggers repair of injured tissues leading to healing.

Since ancient times, inflammation is recognized as a clinical situation, which needs medical attention and care of patients. In fact, the clinical manifestation (known as "the cardinal signs") of inflammation was first described by Aulus Cornelius Celsus, a Roman medical writer who lived in the 1st century AD, as "Rubor et Tumor cum Calore et Dolore" (in Latin) or Redness and Swelling with Heat and Pain (in English) [12].

Depending on the duration of the process, inflammation is classified as either "acute" or "chronic", lasting either for a few hours and days or for weeks and months, respectively. Neutrophils are abundant in acute inflammation, but macrophages and T-lymphocytes are prevalent in chronic inflammation [13]. For this reason, predominant presence of either one of the aforementioned leukocytes in histological examination of tissues is evidence for identifying the respective stage of inflammation.

2.5.2 Inflammation in the Tissue Wound Healing Process

The outcomes of the thrombosis and fibrinolysis processes leave the original wound site full of many bioactive chemical compounds exuded from the injured tissues, released from activated platelets during the various steps of the blood plasma coagulation cascade (Figure 2.6), and produced during the process of fibrinolysis.

In addition, the wound site contains cellular debris composed of dead blood cells (e.g., platelets and red blood cells) as well as dead cells from injured tissues. At this point, the physiological, tissue wound healing process proceeds via inflammation, a stage of paramount importance because, *without inflammation, successful completion of the tissue healing process is impossible.*

2.5.3 The Blood Cells of Inflammation

The inflammatory phase occurs within 24–48 hours post-injury of vascularized tissues and is activated by, and follows (as well as partially overlaps with), the processes of Thrombosis and Fibrinolysis. The environment, rich in bioactive chemicals generated during the Thrombosis and Fibrinolysis processes activates leukocytes, specifically neutrophils and monocytes/macrophages (Table 2.3), major participating blood cells in Inflammation. These cells are chemoattracted at the wound site by bioactive chemical compounds (known as "inflammatory mediators"), including growth factors (e.g., tissue growth factor beta), cytokines (e.g., interleukins), vasodilators (e.g., nitric oxide), etc., which are released and/or produced during the preceding phases of the tissue wound healing process. The mandatory presence of leukocytes at the site of injury necessitates their migration outside the circulation across the blood-vessel wall tissue "vasculature" and chemotactically across various surrounding tissues in order to arrive at the site of tissue injury; this process is known as "extravasation".

2.5.4 Leukocyte Response to Invading Pathogens

The functions of neutrophils and monocytes/macrophages during Inflammation enable the defense of the human body against invading, disease-causing pathogenic microorganisms.

When pathogens, such as bacteria, viruses, fungi, and parasites, invade the human body, they find a hospitable environment with excellent conditions (specifically the appropriate nutrients, temperature, pH, etc.) to survive, grow in numbers, and develop a parasitic existence. The presence and growth of pathogens, either locally or systemically (by subsequent spreading from the original anatomical site of invasion to other parts of the host body), can cause diseases that complicate human welfare and endanger life. Defending the human body, neutrophils (the most abundant subgroup of human leukocytes and the first responders in the process of Inflammation) and monocytes/macrophages (Figure 2.4) respond and annihilate pathogens at the anatomical location of their presence and growth.

This outcome requires leukocyte-pathogen interactions, which are modulated by opsonization. *Opsonins*, including immunoglobulins G (IgG) and M (IgM) from the lymphatic liquid, and molecules of the complement system (specifically C3b) from the circulating blood plasma, are adsorbed onto the membranes of the invading pathogens. Although at this stage the pathogens are still alive, opsonization marks them first for recognition by activated leukocytes known as "phagocytes" and then for subsequent destruction and elimination through phagocytosis (that is, engulfment into the leukocyte cytoplasm followed by "bombardment" with lethal bioactive chemical compounds leading to pathogen destruction).

2.5.5 Neutrophil Response to Invading Pathogens

Because the human body is extensively vascularized, capillaries (the smallest blood vessels in the human body with internal diameters of 5–10 μm and vascular wall tissue composed of an endothelial cell monolayer and a basement membrane) provide the connection between arterioles and venules and, thus, close the blood circulation loop and ensure uninterrupted blood transport in all tissues, organs, and parts of the body.

In the human blood circulation system, neutrophils, the most abundant leukocytes (Table 2.3), are distributed into two "pools": one contains neutrophils circulating in the bloodstream, and the other contains "marginating" neutrophils, which are not activated and are present inside the cardiovascular system close to the nonactivated endothelium (the innermost lining of the blood vessels). From this strategic location, neutrophils (the first responding leukocytes in inflammation) are ready to defend the human body and assure its welfare by killing invading pathogens.

Consider the case of the neutrophil response to a bacterium, which has penetrated the human skin and lodged itself among the muscles. As the pathogen invader survives and replicates in the hospitable environment of the body, it releases chemical compounds (e.g., metabolic waste), which accumulate at the location of the injured, inflamed tissue, affect the vascular tone, and cause irritation to the surrounding tissues. As the pathogen colony grows, these spots may become sites of infection. Concomitantly, opsonins (contained in blood plasma) leak from dilated capillaries and interact with respective receptors on the pathogen membrane, marking it for recognition and subsequent interaction with activated neutrophils and elimination. In addition, the surrounding damaged tissues release pro-inflammatory chemical compounds (such as chemotactic agents and cytokines).

The aforementioned secreted bioactive chemical compounds orchestrate neutrophil activation and several functions, including recruitment from inside the blood vessels, migration through the vascular wall tissue into the surrounding tissues, and arrival at the sites of tissue damage, infection, or inflammation inside and outside of the vascular system. This multistep immune response is unique because it starts while neutrophils are still inside the bloodstream (in a nonactivated or resting state) but continues when "activated" neutrophils (and monocytes) exit the vasculature and migrate across tissues (outside the vasculature), reaching the sites of inflamed/damaged tissues throughout the human body.

The activated neutrophil migration from the bloodstream to the inflamed/infected tissues is a complex, multistep process involving *capturing/tethering*, *rolling*, and firm *adhesion* (while the neutrophils are still inside the vasculature), *diapedesis* (extravasation through the vascular wall tissue when the neutrophils leave the circulation), *chemotaxis* (directed migration in response to a chemical gradient and through various tissues outside the vasculature), and *phagocytosis* (the final step of the neutrophil–pathogen interaction resulting in killing the invader pathogen). Briefly, interactions of receptors on both activated neutrophils and vascular endothelium, as well as the presence of signaling molecules, result in capture of neutrophils on the endothelium (capture/tethering). The dynamic forces of flowing blood initially overcome the contact strength between the aforementioned two activated cells and cause rolling of the neutrophils on the endothelium layer. Eventually, the rolling stops, and the neutrophils adhere (adhesion), slowly crawl on the activated vascular endothelium in search of a passage site between two adjacent endothelial cells (*tight junctions*), squeeze through them and through the basement membrane and thus leave the interior of the blood vessel through transendothelial migration (diapedesis, an irreversible process). Remarkable aspects of this process are the following: completion of extravasation *without damage either to the migrating neutrophils or to the vascular endothelium, or to the surrounding tissues, and no blood leakage*!

Once outside the vasculature, the neutrophil transmigrates through gaps in the pericytes (contractile cells surrounding the outside layer of blood microvessels) and then moves – a process of guided cell migration known as "chemotaxis" – through tissues guided by extracellular chemoattractant chemical gradients directly toward the site of higher concentrations of such chemicals emanated from the live bacterium at the location of the invading pathogen [14, 15]. At the site of the lodged bacterium, the neutrophil recognizes the opsonized bacterium, completely engulfs it through newly (and for this purpose) formed cytoplasmic extensions, and "bombards" the trapped pathogen with various intracellular chemicals (i.e., contents of the neutrophil cytoplasm granules). These potent bioactive chemicals include enzymes (such as collagenase and elastase), oxygen-derived free radicals, oxygen peroxide, etc. The barrage of these potent bioactive chemicals results in killing the invading bacterium. Constituent components of the dead bacterium are eliminated through established physiological procedures. The latter stage of the neutrophil response to invading bacteria is known as "phagocytosis" (Figure 2.8). It should be noted that phagocytosed cellular debris (including dead platelets, neutrophils, red blood cells (erythrocytes), damaged/dead cells from various tissues, etc.) are removed from the human body through normal elimination and secretion processes. The various steps of neutrophil migration from the bloodstream through the vascular wall tissue [16] are similar to that of monocytes (shown schematically in Figure 2.9A) [17].

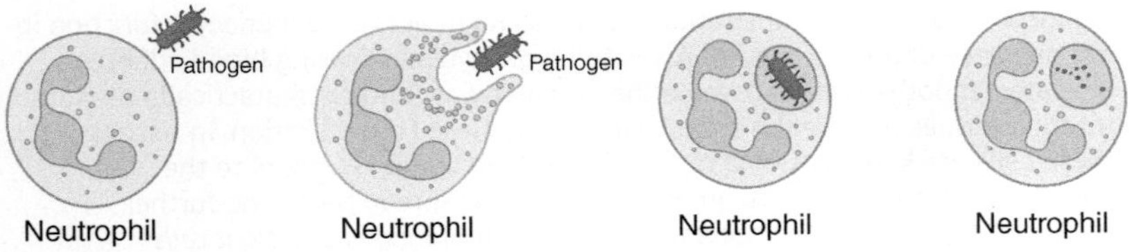

Figure 2.8 Schematic illustration (not to scale) of the phagocytosis process resulting in elimination of a pathogen by an activated neutrophil.

2.5.6 Monocyte/Macrophage Response to Invading Pathogens

Monocytes circulating in the bloodstream respond to invading pathogens via processes similar to the ones used by neutrophils [17]. The main differences in the responses of the two leukocyte types are (i) timing of arrival at the injured/infected site (specifically hours to a few days for neutrophils versus several days for monocytes); and (ii) following migration from the bloodstream into tissues (Figure 2.9A), monocytes differentiate into tissue-resident *macrophages* (Figure 2.9B). Depending on whether macrophages either promote or decrease inflammation (and promote tissue repair), these cells are classified as either "M1" or "M2", respectively.

The principal function of macrophages is phagocytosis, that is, engulfment of pathogens into the macrophage cytoplasm followed by pathogen destruction through intracellular digestion and thus elimination of not only pathogens but also damaged/dead cells (including platelets, neutrophils, erythrocytes, and damaged tissue debris). The various steps of monocyte migration from the bloodstream through the vascular wall tissue and the function of macrophages at the damaged/inflamed tissue sites are shown schematically in Figure 2.9A–C.

Within a few days, activated macrophages and neutrophils at the injured/inflamed tissue site release bioactive chemical compounds (such as growth factors and cytokines), which promote functions (i.e., migration, proliferation, etc.) of the cells of the surrounding intact tissue(s) pertinent to new tissue formation and, thus, to tissue wound healing.

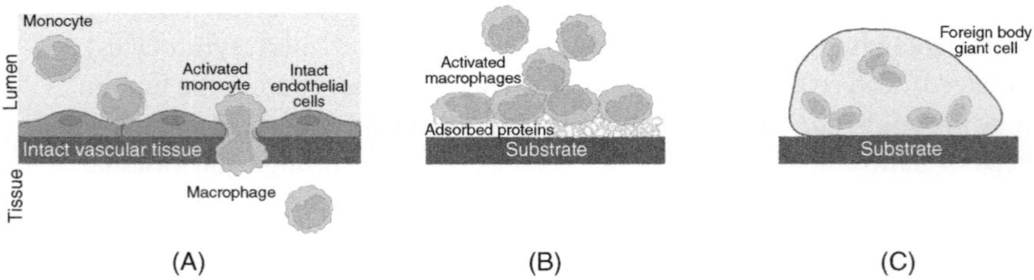

Figure 2.9 Schematic illustration (not to scale) of activated monocyte function in the presence of a substrate. (A) Activated monocytes in flowing blood adhere to activated endothelium, transverse the vascular wall, and chemotactically migrate through tissues outside the vasculature to the site of implantation. In tissues, these cells are known as "macrophages". (B) Macrophages adhere to the pre-adsorbed proteins at the implant surface and spread becoming further activated. (C) Several macrophages fuse and form *foreign body giant cells* characterized by their large size and presence of many nuclei.

2.6 The Proliferation/Repair Stage of the Tissue Wound Healing Process

The events of the inflammatory stage leave the original tissue wound site with a "gap/discontinuity" but free of dead cells and tissue debris, as well as rich in bioactive chemical compounds released during the preceding stages (mentioned in Section 2.2) of the wound healing process. During the Proliferation/Repair stage, tissue continuity is reestablished within 4 days and may continue for up to 21 days after injury [18].

During the Proliferation/Repair stage, the center of action is outside the blood vessels; the focus of these events is on the surrounding, intact tissue, specifically the constituent, anchorage-dependent cells whose functions are programmed to form new tissue. Successful tissue wound repair involves (i) cell regeneration, (ii) synthesis and deposition of new extracellular matrix, (iii) formation of new blood vessels, and (iv) restoration of nerves (mentioned in Chapter 2, Table 2.1). Pertinent responses by these cells are initiated and modulated by select bioactive cytokines and growth factors, many of which have been released at preceding stages of the tissue wound healing process [18].

2.6.1 Cytokines and Growth Factors

Cytokines are small proteins secreted both by several activated leukocytes (i.e., macrophages, B-lymphocytes, T-lymphocytes, and mast cells) and by tissue cells (e.g., endothelial cells and fibroblasts). This group of bioactive chemical compounds includes chemokines, interferons, interleukins, lymphokines, and tumor necrosis factor, which are important signaling molecules mediating and regulating immunity, inflammation, and hematopoiesis in health and disease.

Growth factors are protein signaling molecules which stimulate various cell functions such as differentiation, migration, proliferation, and synthesis and deposition of extracellular matrix components. Some growth factors modulate functions of only specific cells: e.g., vascular endothelial growth factor (VEGF) affects endothelial cell function pertinent to new blood vessel formation (*angiogenesis*), while nerve growth factor regulates axonal growth and branching of nerves, and bone morphogenetic proteins (BMPs) modulate functions of cells in bone and cartilage. In contrast, fibroblast growth factor (FGF) – so named because it was first discovered in promoting fibroblast proliferation – affects functions (such as migration, proliferation, and differentiation) of many other cell types involved in important processes (such as wound healing, angiogenesis, and embryonic development). Platelet-derived growth factor (PDGF) is released from activated platelets and modulates tissue cell (such as fibroblasts and endothelial cells) functions in tissue repair and regeneration. *In vivo*, various growth factors secreted by cells reach their target cells either by diffusion (*paracrine*) or by release into, and transport via blood flow (*endocrine*). In the case of *autocrine* effects, growth factors affect the same cell from which they were released. Many growth factors were first discovered and isolated from nature but are now produced using synthetic biochemical processes in pharmaceutical companies.

2.6.2 Tissue Regeneration

Tissue regeneration and repair require pertinent functions of cells, the living components of tissues. In the case of the Proliferative/Repair stage of the wound healing process [19], these cells are present in the intact tissues surrounding the original wound site, which is now clean and free from infection (a condition resulting from successful completion of the preceding Inflammation stage), and are then primed for subsequent functions pertinent to new tissue formation.

Fibroblasts, ubiquitous cells present in a quiescent stage in tissues throughout the body, are recruited by cytokines and growth factors released by macrophages and platelets during the blood coagulation and inflammatory stages. In response to platelet-derived growth factor, a potent chemotactic chemical, fibroblasts migrate away from the surrounding tissues on the perimeter of the original wound toward the center of the now clean site. Other bioactive chemicals produced at the wound site during the preceding stages of the tissue wound healing process induce fibroblast proliferation within three to five days following tissue injury. At this time, fibroblasts are the predominant cell type present in clean, noninfected wounds. The constituent cells of the surrounding intact tissues begin to participate by migrating, proliferating, and producing the components of the extracellular matrix appropriate for each surrounding tissue.

2.6.3 The Proliferative Capacity of Human Cells

Success of the Proliferation/Repair stage of the wound healing process is determined by the proliferative capacity of constituent tissue cells. In this respect, the cells of human tissues are not created equal.

The proliferative capacity of the cells in the surrounding tissues controls the desirable outcome of the Proliferation/Repair stage of the tissue wound healing process [19]. Limited proliferative capacity of constituent cells is responsible for the lack of regeneration of tissues and organs in humans. Among human organs, *only* the liver has regenerative capacity, specifically, starting from 25% of its original mass, the human liver can grow back to full size. Moreover, among human tissues, very small cuts in skin and small fractures in bone heal.

The reason for the observed variability in tissue and organ regeneration in humans is the very different regenerative/proliferative capacity of constituent cells. Some cells, which are known as "labile", proliferate throughout life without humans being aware that this process is taking place. Examples of such cells include epithelial cells, lymphoid cells, hematopoietic cells in the bone marrow, and gastrointestinal cells. "Stable or expanding" are cells which retain their proliferative capacity, but they do not proliferate without an appropriate stimulus; these cells need either a biochemical or a biophysical stimulus in order to proliferate. Examples of such cells are most of the cells in the tissues of the human body, including liver, kidneys, the ubiquitous fibroblasts, smooth muscle cells, osteoblasts, chondrocytes, vascular endothelial cells, etc. "Permanent or static" are the cells that cannot reproduce themselves after birth. Examples of such cells are nerves, skeletal muscles, and cardiac muscle cells.

It should be noted that many permanent/static cells are constituents of very important organs in the human body. Moreover, it is well known that when these organs (e.g., heart and the central nervous system) are injured, their recovery is problematic, and healing is not achieved; such conditions result in many clinical, pathological conditions which cannot be cured using the current medical armamentarium.

A recent development of great hope for curing the aforementioned injured tissues and organs is stem cells. Since these cells (i) can renew themselves for long periods of time and (ii) have the capacity of becoming cells of all other tissues in the body, it is hoped that they can be used to replace the nonproliferative cells of the heart, nerves, and skeletal muscles of patients suffering from diseases which have been difficult (if not impossible) to cure so far. Treating patients with stem cells is a medical field in development. Stem cells definitely have tremendous potential and promise for helping patients.

2.6.3.1 An Introduction to Stem Cells

Stem cells are unspecialized populations of cells which divide, renew over long periods of time, and upon receiving appropriate stimulation, they differentiate and give rise to specialized cell constituents of all human tissues. Stem cell *potency* refers to their ability to differentiate. The hierarchy of stem cell potency includes totipotent, pluripotent, multipotent, unipotent, and progenitor stem cells. Totipotent refers to the ability of stem cells to form an organism (e.g., for humans, totipotent cells give rise to the human embryo, the placenta, and additional embryonic tissues). Pluripotent stem cells can only differentiate in the embryo into three germ layers: endoderm, mesoderm, and ectoderm. An example of pluripotent stem cells is embryonic stem cells and induced pluripotent stem cells. Multipotent stem cells can differentiate into more than one type of specialized differentiated cells; hematopoietic stem cells (HSCs) and mesenchymal stem/stromal cells (MSCs) are pertinent examples. Unipotent stem cells differentiate into only one type of specialized cell such as muscle stem cells. Progenitor cells are intermediate cells between stem cells and differentiated cells; since these cells are not fully differentiated, they are not specialized regarding cell function. Progenitor cells differentiate into a limited number of specialized cells and have reduced capacity for self-renewal; a pertinent example is endothelial progenitor cells [20].

Embryonic Stem Cells Embryonic stem cells are pluripotent stem cells isolated from the *blastocyst*, a cluster of cells formed in the inner cell mass during early embryonic development in mammals [20]. Because embryonic stem cells are pluripotent, they can differentiate into cells of any of the tissues in the three germ layers: ectoderm, mesoderm, and endoderm. The ectoderm directs formation of the epidermis, nervous system, etc. The mesoderm gives rise to cells of the blood circulation system, skeletal system, reproductive system, etc. The endoderm leads to formation of epithelial cell linings of the digestive tract and respiratory systems, pancreas, thyroid, etc. Embryonic stem cells can proliferate indefinitely and thus have vast potential in, and are highly sought after for, research and subsequent biomedical applications. However, the full potential of embryonic stem cells has not been fully explored and remains partially understood. Engineers, biologists, and other researchers have been investigating the use of embryonic stem cells to address biomedical-related questions.

Induced Pluripotent Stem Cells Induced pluripotent stem cells (iPSCs) are generated from somatic cells, a differentiated cell type. iPSCs are generally reprogrammed from either differentiated skin cells or blood cells, isolated from either a donor or patient directly and reverted back to a pluripotent cell. A set of reprogramming parameters are required to expose the somatic cell in order to revert this cell to the pluripotent stage. Such pioneering research was awarded the Nobel Prize in Physiology or Medicine in 2012 [21].

Details of the four specific transcription factors (specifically Oct 3/4, Sox2, c-Myc, and Klf4) to revert somatic cells to pluripotent cells were reported in a seminal article published in 2006 [22]. In a follow-up to this publication, the same research group detailed the process to obtain iPSCs from primary human dermal fibroblasts and the methods used to confirm cell pluripotency [23]. While both iPSCs and embryonic stem cells are pluripotent, advantages of using iPSCs over embryonic stem cells include the ability to obtain these cells from somatic cells, their relevance for disease modeling (i.e., these cells can be obtained from an individual with a specific disease) and subsequent utilization for personalized medical treatment.

Adult Stem Cells Adult stem cells are either multipotent or unipotent, have the capacity for self-renewal, and give rise to other mature cell types (at the final differentiation stage) [20]. Because these cells are mature cells, they have specific morphologies and cell functions that enable them to carry out specific functions of the tissues in which they reside; for these reasons, the primary function of adult stem cells is to regulate tissue homeostasis (for example, in response to tissue injury or to replenish cells after apoptosis/necrosis). It is important to note that human adults have a small number of stem cells, which decrease with age.

Examples of adult stem cells include HSCs and MSCs; both are harvested from the bone marrow and are characterized by select positive and negative markers, providing identification of specific cell types. HSCs, multipotent stem cells with the ability to self-renew, were first isolated from the red bone marrow in the flat bones and also from umbilical cord blood and peripheral blood of adults [24]. HSCs are involved in the process of hematopoiesis (referred in Chapter 2, Section 2.2.3), and give rise to two types of progenitor cells: myeloid progenitor cells and lymphoid progenitor cells. The myeloid progenitor cells differentiate into red blood cells, platelets, and white blood cells; the lymphoid progenitor cells differentiate into T cells.

MSCs, are multipotent stem cells with proliferative capacity and the ability of self-renewal, were first isolated from the yellow bone marrow of long bones and also from many other tissues, such as umbilical cord blood, adipose tissue, placental, etc. MSCs have the ability to differentiate into osteoblasts, chondrocytes, adipocytes, myoblasts, and tenocytes. Due to their ease of isolation, proliferative capacity, and multipotency, MSCs have been widely explored for use in various cell-based medical therapies [25].

Stem-Cell-Based Therapies Stem-cell-based therapies require transplantation of human stem cells to repair or replace damaged cells and/or tissues. The most well-known stem cell therapy used to date is hematopoietic stem cell transplantation (also referred to as "the bone marrow transplant"), used to treat a variety of blood cancers and hematologic pathologies. While other cell types may be used for cell therapies, stem cells are most promising due to their capability to self-renew, maintain stemness, and differentiate. For these reasons, stem cell therapies hold promise in chronic disease applications such as cancer, autoimmune diseases, musculoskeletal diseases, spinal cord injuries, etc. Due to their ease of isolation, proliferative capacity, and multipotency, MSCs have been explored for use in cell-based therapies [25]. While the number of clinical trials documented in the United States using MSCs has increased annually, the percentage of Phase III clinical trials, however, was only 6% in the decade 2010–2020 [25]. The slow progress is due to unresolved challenges using stem cells for cell-based therapies, which include the following: large (1–200 million/dose) cell numbers are required for each patient treatment, cells must be expanded ex vivo, and direct injection or systemic delivery of cells results in low cell survival and retention following delivery [25].

In order to meet the extremely large numbers of stem cells required for medical therapeutic applications for each patient, biomedical engineering endeavors have focused on developing techniques and instrumentation to increase the *in vitro* numbers of undifferentiated stem cells for clinical applications. In this respect, the pertinent focus is to (i) *maintain stemness*, i.e., to maintain stem cells in an undifferentiated state and (ii) prevent unwanted and uncontrolled differentiation of stem cells into several cell types. For example, in the case of MSCs, preventing uncontrolled/undesirable differentiation translates to obtaining exclusive MSC differentiation into only one cell type, for example, osteoblasts (without differentiation into either chondrocytes or adipocytes).

Organoids Another biomedical application of stem cells is organoids, i.e., three-dimensional cellular structures derived from either pluripotent or multipotent stem cells. These constructs have gained attention in the biomedical research field because they can be used as model human systems to investigate many biological processes without use of animals. Over two-dimensional cellular models, organoids have advantages for modeling human physiology due to the capability of stem cells to self-organize into three-dimensional tissue structures *in vitro*. These tissue models are a unique biological category, replicating both tissues and organs, recapitulating many biological parameters, such as spatial organization and physiological functions of tissue-specific cells, in addition to cell–cell and cell–matrix interactions of both undifferentiated and differentiated cells. Organoids bridge the gaps between existing model systems for research endeavors by providing a representative mammalian, controlled environment responsive to select modulations/stimuli.

Organoids have been used to investigate the processes involved in disease initiation and progression, for drug discovery applications, in high throughput screening methods, as well as tissue alternatives for regenerative medicine applications. Although many stem-cell-derived organoids have been developed, such systems involve several, but not all, tissue components [26]. For this reason, engineers and scientists strive to investigate the conditions which direct cell functions and organization (the vital processes needed for organoid formation) while elucidating the underlying mechanism(s) responsible for such outcomes. Using multidisciplinary approaches, organoids have been constructed and deconstructed to provide enhanced control of organoid formation for specific applications [26]. For this purpose, the process of organoid formation is initiated under cell culture conditions using either pluripotent stem cells or multipotent adult stem cells. Under appropriate biophysical and biochemical cues (comparable to those in the specific *in vivo* environment), cell aggregates form and then are embedded in a hydrogel matrix (most commonly, Matrigel). When cultured in the presence of appropriate niche chemical components, the cells self-assemble, differentiate, and generate organoids composed of several cell types. The resultant organoids can be used to recapitulate organogenesis, serve as systems for disease modeling, high throughput drug testing, as well as methods for studying mammalian development [26].

Single-Cell Omics Omics pertain to the analysis of large "volumes" of structure and function data representing the entire makeup of biological systems (i.e., genomics, transcriptomics, epigenomics, proteomics, and multiomics) [27]. Single-cell sequencing methods provide insight at the highest resolution of stem cell heterogeneity; such methods represent a powerful technology that enables analysis of large numbers of cells to characterize rare cell types and understand various cell states [28, 29]. Single-cell omics can be applied to developmental studies, precision medicine approaches in humans, and the Human Cell Atlas, a global consortium mapping every cell type in the human body.

The potential of omics has only started to be explored. The promise of advances and contributions of this field in the future is great.

2.6.4 Synthesis and Deposition of New Extracellular Matrix Components

In addition to cell regeneration, tissue repair after injury requires formation of a new extracellular matrix. This process starts with contributions from fibroblasts; these cells respond to the various bioactive chemical stimuli at the site of healing tissues by synthesizing and depositing chemical compounds (such as collagens, fibrin, fibronectin, proteoglycans, glycosaminoglycans, etc.) [30]. The deposited chemical compounds form a *provisional extracellular matrix*, which becomes the platform for subsequent migration of the constituent primary cells from the surrounding intact tissue. This stage of tissue repair is known as "fibroplasia" [31].

In addition to migration over the provisional extracellular matrix toward the center of the original wound site, the bioactive chemicals (e.g., growth factors) released at the original wound site during the preceding stages modulate constituent cells in the surrounding tissues, promoting functions (such as cell proliferation, synthesis and deposition of extracellular matrix components, etc.) pertinent to new tissue formation. The constituent cells of each tissue (for example, osteoblasts in bone, endothelial cells, and smooth muscle cells in the vasculature, epithelial cells in skin) are programmed by nature to accomplish this task.

The deposited components of the extracellular matrix are specific and characteristic chemical compounds of each tissue. Both the type and relative amount of the chemical compounds present in the extracellular matrix are different in each tissue. For example, the extracellular matrix of bone contains both Collagen Type I and non-collagenous proteins (specifically osteonectin, osteocalcin, osteopontin, proteoglycans, and glycosaminoglycans), the vascular wall tissue contains laminin, elastin, several types of collagen, and fibronectin, and skin contains collagens, elastin, proteoglycans, fibronectin, etc. [31]. The overall purpose of the extracellular matrix components is to maintain tissue structure and to retain tissue function.

Collagen is present in many tissues of the human body. To date, 28 types of collagen have been identified. Examples of known types of collagen include Type I (present in bone, skin, tendon, and ligaments), Type II (in cartilage), Type IV (in basal lamina, a component of the basement membrane of the vascular wall tissue), and Type VII (found in skin and fetal membranes). Various cells produce different collagen types: for example, osteoblasts synthesize Collagen Type I, chondrocytes synthesize Collagen Type II, and fibroblasts synthesize Collagen Type III [32]. Collagen synthesis begins 48 hours post tissue injury; it is deposited and accumulates in the extracellular matrix of tissues during the Proliferation/Repair stage of the tissue wound healing process. The presence, type, and relative amount of collagen contribute to the tensile strength of tissues.

Other important bioactive chemical compounds for the Proliferation/Repair stage of the tissue wound healing process are growth factors (discussed in Chapter 2, Section 2.6.1). Various types of growth factors are naturally present in the human body in health and in disease; they are synthesized and/or released from activated blood cells and from tissue cells during tissue injury as well as during the Blood Coagulation, Fibrinolysis and Inflammation stages of the tissue wound healing process (discussed in Chapter 2, Section 2.3, Section 2.4, Section 2.5, respectively). Once released, growth factors are sequestered in the extracellular matrix of tissues can be activated and thus become available to modulate pertinent cell functions (such as proliferation, etc.). In terms of bioactivity, some growth factors are bioactive only when they are adsorbed and attached to chemical components of the extracellular matrix; other growth factors are bioactive only when, after release, they are dissolved in biological liquids.

2.6.5 Formation of New Blood Vessels

Very few tissues in the human body, including cartilage, epithelial cells of the skin, and cornea of the eye, are avascular; all other tissues are vascularized. This aspect reflects the critical fact that survival and function of tissues and organs in the human body require supply of oxygen and nutrients, as well as removal of carbon dioxide and cell metabolic waste. For this reason, and in order to survive, cells must reside within 150–200 μm from the blood supply provided by capillaries [33].

The aforementioned requirement is attained by formation of new blood vessels (a process known as "angiogenesis") during the Proliferation/Repair stage. This outcome is critical and requisite for successful completion of the wound healing process, formation of new tissue, and subsequent long-term survival and normal function of tissues and organs [33].

At the capillary level, formation of new blood vessels results from migration and proliferation of endothelial cells, as well as from synthesis and deposition of chemical compounds forming a basal membrane by these cells. Pertinent endothelial cell functions are stimulated and promoted by various growth factors (such as platelet-derived growth factor, basic FGF, insulin-like growth factor one, etc.) present at the site of tissue injury and produced during the preceding stages. Another important stimulus for new blood vessel formation is hypoxia, a condition resulting from the destruction of blood vessels during tissue injury and dominant during the Blood Coagulation and Inflammation stages of the tissue wound healing process.

The large number of forming capillaries at the skin tissue wound site during the Proliferation/Repair stage of the healing process results in the formation of new *granulation tissue*, which is red in color (due to red blood cells inside capillaries) with an irregular, rough surface. Histologically, granulation tissue contains numerous blood vessels, inflammatory cells, fibroblasts, and a minimal amount of mature collagen. The presence of an extracellular matrix (which is not "compact" or dense yet) is evidence that new tissue is in the process of being formed [34].

2.6.6 Restoration of Nerves

Proper function of tissues, organs, body parts, and, in fact, the entire human body requires a healthy nervous system. For this reason, successful completion of the wound healing process requires restoration of nerves damaged during injury in order to obtain functional tissues, organs, and bodies.

Recognition of the fundamental importance of this aspect of tissue wound healing has resulted in pertinent research endeavors at the cellular-, molecular- and gene-levels enabled by recent advances in various scientific fields including cellular/molecular biology, biochemistry, tissue engineering, biomaterials, and bioinstrumentation. Undoubtedly, this is a cutting-edge area of scientific inquiry with great potential and promise for seminal contributions in the future.

2.7 Remodeling/Maturation Stage of the Tissue Wound Healing Process

The need to repair an injured tissue has priority during the events of the Proliferation/Repair stage of the wound healing process; emphasis is on constituent cell proliferation as well as on the synthesis and deposition of the chemical components of the extracellular matrix of the healed tissue. At the end of the Proliferation/Repair stage, the new tissue is characterized by chemokine suppression, a fibrous matrix, and many immature blood vessels but has not achieved either the cell and chemical compositions or the functional capability of normal, uninjured tissue [19, 35].

Restoration of normal, functional tissues requires attention to details, which are addressed during the Remodeling/Maturation, the last stage of the wound healing process, and include edema loss, wound contraction, maturation of the vasculature, degradation of the provisional extracellular matrix, and, most important, reorganization of the new extracellular matrix evidenced by collagen cross-linking and an "equilibrium" between collagen synthesis and collagen degradation.

Edema, the result of liquid transferred from blood plasma by oozing across the blood vessel wall tissue, accumulating at the site of the inflamed tissues, is one of the cardinal signs of Inflammation (discussed in Chapter 2,

Section 2.5). Edema loss (responsible of the characteristic swelling and puffiness) indicates that edema subsides during the last stage of the tissue wound healing process.

Furthermore, beginning at five days after injury, the clean, noninfected wound contracts because its edges are pulled toward the center of the original injury site as a result of the function of myofibroblasts (which are cells with contractile properties) and components of the extracellular matrix (such as elastin, collagen, etc.), which have pertinent mechanical properties [19]. Moreover, the provisional extracellular matrix degrades. The composition and relative distribution of various chemical compounds in the extracellular matrix of the healed tissues are most important because they mediate, and thus dictate, the subsequent interactions and functions of anchorage-dependent cells. During the Remodeling/Maturation stage of the tissue wound healing process, it is necessary that the predominant cells of a specific tissue acquire the morphology and orientation characteristic of the normal, intact tissue, as well as proliferate, synthesize, and deposit the chemical components of their extracellular matrix; these cell contributions are essential for the formation of new, functional tissue.

During formation of the new extracellular matrix, certain types of collagen may be replaced by others (for example, in the case of skin, Collagen Type III is replaced by Collagen Type I) [36]. Most importantly, the randomly deposited collagen is organized and cross-linked (an outcome which decreases collagen degradation and improves the tensile strength of the restored tissue). It should be noted, however, that while repaired tissues gradually become stronger with time, they never acquire the full mechanical properties of the normal tissue before injury; for example, remodeled skin (the classic model of tissue wound healing) reaches only 80% of the normal/uninjured skin value [37]. Overall, the Remodeling/Maturation stage is a process which refines the outcome of the tissue wound healing process to obtain, as close as possible, normal, functional tissue.

References

1 Kaushansky K, Prchal JT, Burns LJ, Lichtman MA, Marcel Levi DCL, editors. *Williams Hematology*. 10th ed. McGraw Hill; 2021.

2 Goldman L, Cooney KA, editors. *Goldman-Cecil Medicine*. 27th ed. Elsevier; 2023.

3 Hoffman R, Benz EJ, Silberstein LE, Heslop H, Weitz J, Salama ME, Abutalib SA, editors. *Hematology Basic Principles and Practice*. 8th ed. Elsevier; 2022.

4 Lovett M, Lee K, Edwards A, Kaplan DL. Vascularization strategies for tissue engineering. *Tissue Eng. Part B Rev.* [Internet]. 2009 [cited 2022 Mar 16];15(3):353. Available from: https://www.ncbi.nlm.nih.gov/pmc/articles/PMC2817665/.

5 The bone marrow and blood formation - Leukaemia Foundation (2024) https://www.leukaemia.org.au/blood-cancer/understanding-your-blood/bone-marrow-and-blood-formation/ (accessed 09 June 2024).

6 Dzierzak E, Philipsen S. Erythropoiesis: development and differentiation. *Cold Spring Harbor Perspect. Med.* [Internet]. 2013 [cited 2022 Mar 15];3(4). Available from: /pmc/articles/PMC3684002/.

7 Williams Hematology, 10th ed. | AccessMedicine | McGraw Hill Medical [Internet]. [cited 2022 Mar 16]. Available from: https://accessmedicine-mhmedical-com.libproxy.uthscsa.edu/book.aspx?bookid=2962.

8 Mortaz E, Alipoor SD, Adcock IM, Mumby S, Koenderman L. Update on neutrophil function in severe inflammation. *Front. Immunol.* 2018; 9:411381.

9 Kapila V, Wehrle CJ, Tuma F. *Physiology, Spleen*. StatPearls [Internet]. 2021 [cited 2022 Mar 16]; Available from: https://www.ncbi.nlm.nih.gov/books/NBK537307/.

10 Barmore W, Bajwa T, Burns B. *Biochemistry, Clotting Factors*. StatPearls [Internet]. 2021 [cited 2022 Mar 28]; Available from: https://www.ncbi.nlm.nih.gov/books/NBK507850/.

11 Chapin JC, Hajjar KA. Fibrinolysis and the control of blood coagulation. *Blood Rev.* [Internet]. 2015 [cited 2023 Sep 1];29(1):17. Available from: https://www.ncbi.nlm.nih.gov/pmc/articles/PMC4314363/.

12 NLM. Historical Perspectives – Inflammation and the Microcirculation – NCBI Bookshelf [Internet]. [cited 2022 Mar 28]. Available from: https://www.ncbi.nlm.nih.gov/books/NBK53379/.

13 Babensee JE. 2.2.2 – Inflammation, wound healing, the foreign-body response, and alternative tissue responses, Editor(s): William R. Wagner, Shelly E. Sakiyama-Elbert, Guigen Zhang, Michael J. Yaszemski, *Biomaterials Science*. 4th ed. Academic Press, 2020, Pages 737–746, ISBN 9780128161371, https://doi.org/10.1016/B978-0-12-816137-1.00048-9. (https://www.sciencedirect.com/science/article/pii/B9780128161371000489)

14 Choi EY, Santoso S, Chavakis T. Mechanisms of neutrophil transendothelial migration. *Front. Biosci.* [Internet]. 2009 [cited 2023 Sep 1];14(5):1596. Available from: https://pubmed.ncbi.nlm.nih.gov/19273149/.

15 Muller WA. Getting leukocytes to the site of inflammation. *Vet. Pathol.* 2013;50:7–22.

16 Petri B, Sanz MJ. Neutrophil chemotaxis. *Cell Tissue Res.*2018 [cited 2023 Sep 1];371(3):425–36. Available from: https://link.springer.com/article/10.1007/s00441-017-2776-8.

17 Shi C, Pamer EG. Monocyte recruitment during infection and inflammation. *Nat. Rev. Immunol.* [Internet]. 2011 [cited 2022 Mar 28];11(11):762–74. Available from: https://pubmed.ncbi.nlm.nih.gov/21984070/.

18 Landén NX, Li D, Ståhle M. Transition from inflammation to proliferation: a critical step during wound healing. *Cell. Mol. Life Sci.* [Internet]. 2016 [cited 2020 Jul 5];73:3861–85. Available from: https://www.ncbi.nlm.nih.gov/pmc/articles/PMC5021733/.

19 Kordestani SS. Wound Healing Process. Atlas of Wound Healing 2019;11–22.

20 ISSCR. *Learning Guide.* International Society for Stem Cell Research [Internet]. [cited 2024 Apr 2]. Available from: https://www.isscr.org/learning-guide.

21 The Nobel Prize in Physiology or Medicine 2012 – Press Release. [Internet]. [cited 2024 Apr 2]. Available from: https://www.nobelprize.org/prizes/medicine/2012/press-release/.

22 Takahashi K, Yamanaka S. Induction of pluripotent stem cells from mouse embryonic and adult fibroblast cultures by defined factors. *Cell* 2006;126(4):663–76.

23 Takahashi K, Tanabe K, Ohnuki M, Narita M, Ichisaka T, Tomoda K, et al. Induction of pluripotent stem cells from adult human fibroblasts by defined factors. *Cell* [Internet]. 2007 [cited 2024 Mar 29];131(5):861–72. Available from: https://pubmed.ncbi.nlm.nih.gov/18035408/.

24 Lee JY, Hong SH. Hematopoietic stem cells and their roles in tissue regeneration. *Int. J. Stem Cells* [Internet]. 2020 [cited 2024 Mar 30];13(1):1. Available from: https://www.ncbi.nlm.nih.gov/pmc/articles/PMC7119209/.

25 Wechsler ME, Rao VV, Borelli AN, Anseth KS. Engineering the MSC secretome: a hydrogel focused approach. *Adv. Healthc. Mater.* [Internet]. 2021 [cited 2021 Mar 15];2001948. Available from: https://onlinelibrary.wiley.com/doi/10.1002/adhm.202001948.

26 Wechsler ME, Shevchuk M, Peppas NA. Developing a multidisciplinary approach for engineering stem cell organoids. *Ann. Biomed. Eng.* [Internet]. 2020 [cited 2021 Aug 29];48(7):1895–904. Available from: https://pubmed.ncbi.nlm.nih.gov/31659603/.

27 Dai X, Shen L. Advances and trends in omics technology development. *Front. Med.* [Internet]. 2022 [cited 2024 Apr 2];9:911861. Available from: https://www.ncbi.nlm.nih.gov/pmc/articles/PMC9289742/.

28 Stein CM, Weiskirchen R, Damm F, Strzelecka PM. Single-cell omics: overview, analysis, and application in biomedical science. *J. Cell. Biochem.* [Internet]. 2021 [cited 2024 Apr 2];122(11):1571–8. Available from: https://pubmed.ncbi.nlm.nih.gov/34459502/.

29 Kumar P, Tan Y, Cahan P. Understanding development and stem cells using single cell-based analyses of gene expression. *Development* [Internet]. 2017 [cited 2024 Apr 2];144(1):17. Available from: https://www.ncbi.nlm.nih.gov/pmc/articles/PMC5278625/.

30 Tracy LE, Minasian RA, Caterson EJ. Extracellular matrix and dermal fibroblast function in the healing wound. *Adv. Wound Care* https://home.liebertpub.com/wound [Internet]. 2016 [cited 2023 Sep 1];5(3):119–36. Available from: https://www.liebertpub.com/doi/10.1089/wound.2014.0561.

31 Gonzalez ACDO, Andrade ZDA, Costa TF, Medrado ARAP. Wound healing – a literature review. *An. Bras. Dermatol.* [Internet]. 2016 [cited 2023 Sep 1];91(5):614. Available from: https://www.ncbi.nlm.nih.gov/pmc/articles/PMC5087220/.

32 Ricard-Blum S. The Collagen family. *Cold Spring Harbor Perspect. Biol.* [Internet]. 2011 [cited 2023 Sep 1];3(1):1–19. Available from: https://www.ncbi.nlm.nih.gov/pmc/articles/PMC3003457/.

33 Phelps EA, García AJ. Engineering more than a cell: vascularization strategies in tissue engineering. *Curr. Opin. Biotechnol.* 2010;21(5):704–9.

34 Alhajj M, Bansal P, Goyal A. *Physiology, Granulation Tissue*. StatPearls [Internet]. 2022 [cited 2023 Sep 1]; Available from: https://www.ncbi.nlm.nih.gov/books/NBK554402/.

35 Ridiandries A, Tan JTM, Bursill CA. The role of chemokines in wound healing. *Int. J. Mol. Sci.* [Internet]. 2018 [cited 2020 Jul 5];19. Available from: https://www.ncbi.nlm.nih.gov/pmc/articles/PMC6214117/.

36 Singh D, Rai V, Agrawal DK. Regulation of Collagen I and Collagen III in tissue injury and regeneration. *Cardiol. Cardiovasc. Med.* [Internet]. 2023 [cited 2023 Sep 1];7(1):5. Available from: https://www.ncbi.nlm.nih.gov/pmc/articles/PMC9912297/.

37 Ireton JE, Unger JG, Rohrich RJ. The role of wound healing and its everyday application in plastic surgery: a practical perspective and systematic review. *Plast. Reconstr. Surg. Glob. Open* [Internet]. 2013 [cited 2023 Sep 1];1(1). Available from: https://pubmed.ncbi.nlm.nih.gov/25289204/.

3

Lessons Learned from Pathology

The ideal and most desirable outcome in the wound healing process of injured tissues (described in Chapter 2) is not always achieved. Differences arise from pathological complications which jeopardize the welfare, and even endanger the lives, of afflicted patients. Representative cases are reported in the medical literature. With respect to wound healing, it is instructive to review some well-known pathological, clinical complications at each stage of tissue wound healing and compare them with the respective events during the "normal" process (reviewed in Chapter 2).

3.1 Blood-Coagulation-Related Clinical Complications

Blood thrombi are either the cause or the unavoidable complication of several pathological conditions. In the case of blood coagulation, some problems that result in clinical complications are related to platelets and to pathologies of the platelet progenitor cells in the bone marrow. Once platelets are produced, mature in the bone marrow, and are released into the human blood circulation, they may be either too few or too many for the required normal platelet contributions to the blood coagulation process. In other pathological syndromes, the granules in the platelet cytoplasm may be empty, i.e., they do not contain the chemical compounds which, upon their release, induce and contribute to platelet activation and subsequent functions (such as adhesion and aggregation). Moreover, the prescribed medications (such as aspirin and other cyclooxygenase inhibitors, which patients may take for pathologies unrelated to platelet-related conditions) may counteract, impair, and sometimes exacerbate platelet functions, which are critical contributions to blood coagulation.

Concepts of Tissue-Biomaterial Interactions: Fundamentals and New Directions,
First Edition. Rena Bizios and Marissa E. Wechsler.

Companion Website: www.wiley.com/go/Concepts_of_Tissue-Biomaterial_Interaction_1e

Other patients are affected by various coagulation factor pathologies (reviewed in Chapter 2, and illustrated in Figure 2.6). There are several pathological hematology syndromes. An illustrative example is hemophilia A, an inherited, pathological condition which affects mostly men; the women of the patient's families are carriers of hemophilia. Extensive research provided evidence that the underlying etiology of hemophilia A is related to gene-level mutation of Coagulation Factor VIII. In some cases, inadequate amounts of bioactive Coagulation Factor VIII are produced and released into circulating blood; in other cases, adequate amounts of bioinactive Coagulation Factor VIII are produced and released into the human blood circulation. Both hemophilia A cases are associated with unwanted, uncontrolled bleeding because the blood coagulation process in the afflicted patients cannot successfully complete the plasma coagulation cascade and, thus, formation of the fibrin net, stabilization of the platelet plug, and generation of the necessary blood clots (reviewed in Chapter 2 and illustrated in Figure 2.6). The excessive and non-stopping bleeding observed in hemophilia results in clinical problems which endanger both the welfare and lives of hemophilia patients.

Clinical problems also arise as a result of either unwanted or excessive blood clotting in vessels that supply blood to major organs (such as the heart, brain, kidneys, etc.) as well as to all other parts of the human body. Examples of clinical thrombotic complications are heart attacks (resulting from blood clots in the coronary arteries), strokes (major clinical complications in the brain caused by blood clots in the carotid arteries), phlebitis (a condition resulting from clots obstructing blood circulation in the legs), etc.

3.2 Prevention of Blood Clots: A Clinical Perspective

Patients with propensity to blood clot formation are treated with medications which prevent platelet activation; these chemical compounds include aspirin and other commercially available cyclooxygenase inhibitors, which prevent formation of thromboxane A_2 and subsequent platelet aggregation, release of the content of cytoplasmic granules, and contribution to thrombus formation.

Patients with hemophilia A receive transfusion of Coagulation Factor VIII following proper extraction from donated whole human blood using pertinent, advanced hematological methodologies and the latest biomedical instrumentation. Such transfusions administered at critical moments stop unwanted bleeding and improve the hemophiliac patients' quality of life.

3.3 Fibrinolysis: Clinical Practice

Formation of unwanted and uncontrolled blood clots in patients may have serious pathological conditions (such as heart attacks, strokes, etc.): in the brain, lack of blood flow for more than three to four minutes [1] causes irreversible damage of brain tissue; in the heart, various regions of the myocardium are damaged irreversibly when deprived of oxygenated blood. In order to avoid such detrimental outcomes, clinical treatments use specially formulated chemical compounds to dissolve thrombi as close to their formation as possible. Fibrinolytic agents (such as tissue plasminogen activator) are available medications which are administered by health-care personnel to break down obstructing blood clots and, thus, restore blood flow in the affected region(s) of the patients' bodies.

3.4 Inflammation-Related Pathological Outcomes

Clinical complications also arise during Inflammation, the next stage of the wound healing process (reviewed in Chapter 2). In this case, pertinent pathological conditions involve the white blood cells, the protagonists of inflammation. Some diseases affect the number and function of pertinent stem cells and progenitor cells in the bone marrow and, consequently, the number of the white blood cells produced and released into the blood circulation. Other pathological conditions have the appropriate numbers and types of white blood cells but are functionally impaired cells.

When peripheral blood is drawn (e.g., during medical checkups) and subsequently analyzed in hematology laboratories, the results provide the numbers of each circulating white blood cell type. The attending medical personnel use this information to determine the status of each patient's health condition as well as to diagnose pathological syndromes. Invariably, high numbers of circulating white blood cells are evidence that the human body is fighting a disease, for example, one caused by invading pathogens (e.g., associated with the common cold) or more serious diseases such as cancer (e.g., leukemia). Proper diagnosis of each pathological condition is followed by pertinent prescribed medication(s) to treat and cure patients.

Overall, special attention to, and consideration for, inflammation is required for both the beneficial and adverse potential results of this stage of the tissue wound healing process. First and foremost, inflammation is imperative because without it, tissue wound healing cannot be achieved. Inflammation-related clinical complications, however, arise when inflammation persists and, from an acute-and-local, become a chronic and systemic condition – in other words, when inflammation spreads to many other parts and organs of the body and lasts for long periods of time. Extensive, uncontrolled chronic inflammation (either local or systemic) is undesirable because it may trigger activation of autoimmune syndromes and may cause rejection of transplanted organs and implanted medical devices.

3.5 Proliferation/Repair and Remodeling/Maturation-Related Pathological Outcomes

The pathological outcomes during the Proliferation/Repair and Remodeling/Maturation stages of the tissue wound healing process are illustrated with a short review of the healing of skin injuries. Skin, the quintessential model in this field, has been extensively studied in terms of its wound healing capabilities and desirable outcomes, as well as undesirable complications.

3.5.1 The Skin Wound Healing Process

The process of wound healing in skin occurs within 48 hours and lasts for up to two to three weeks after injury [2]. In all cases, injury damages *epithelial cells* on the outermost skin layer. During the Proliferation/Repair stage, epithelial cells from the surrounding intact skin flatten and migrate in the horizontal direction toward the center of clean, noninfected wound sites. At this point, cell shape and morphology are adjusted as needed in order to cover the gap of missing tissue at the wound site. Epithelial cells multiply via mitosis, pile up in the vertical direction, recover their characteristic shape and morphology, and gradually cover the entire wound site. This process is called either "epithelialization" or "re-epithelialization". The process of new skin tissue formation is completed by synthesis and deposition of the appropriate extracellular matrix chemical components. An important characteristic component of the skin extracellular matrix is Collagen Type III, which is deposited in thin, randomly oriented strands during the Proliferation/Repair stage of the wound healing process [2].

The important and desirable outcome during the Remodeling/Maturation stage in the skin wound healing process is to regain tissue function. What is needed, therefore, is to gradually obtain new tissue, which exhibits the mechanical properties of healthy, intact skin. During the healing process, the skin tensile strength increases for up to eight weeks after injury [3]. This outcome is correlated with the type and amount of collagen deposited during the Proliferation/Repair stage. During the Remodeling/Maturation stage, appropriate collagen alignment and linking take place in order to obtain functional, stronger, and healed skin. The tensile strength of the healing skin increases slowly with time for up to one year (and may continue for up to two years) after injury [3]. At best, however, the tensile strength of healed tissue reaches only 80% that of the normal (non-wounded) skin [4].

Extensive biomedical research has revealed that the anatomical location, type, depth, and wound size affect the skin wound healing process and outcomes. A development which complicates the progress and outcome of the healing process in many tissues (including skin) is infection. Long-term (i.e., chronic) infection is undesirable because it causes clinical complications requiring special medical treatment of inflicted patients. Other contributing aspects are patient age and health status. In fact, skin heals better during the early, but not during the latter, years of human life. Patient health is always a critical parameter in achieving the desirable outcome in healing many injured tissues, including skin.

3.5.2 Undesirable Skin Wound Healing Outcomes: Scars

The undesirable skin wound healing outcomes are "scars". Scars are formed during skin wound repair in humans of all ages after birth. Similar to other tissues, the early scar has a red color due to many blood-containing capillaries at the tissue injury site during the Proliferation/Repair stage of the wound healing process. During the Remodeling/Repair stage (which in skin takes place over months to two years following injury), the scar is mature and usually hypopigmented (i.e., lighter in pigmentation than the surrounding intact skin). In addition, skin scars do not have hair follicles and glands. In terms of histology, the collagen in scars is present as dense, packed fibers; this kind of collagen arrangement is not present in normal, unwounded skin [5].

Clinical examples of skin scars are (i) hypertrophic scars and (ii) keloids. Both scar types are esthetically undesirable outcomes when they are formed on visible parts (such as the face, neck, ears, arms, etc.) of the bodies of patients. Hypertrophic scars are located within the boundaries of the original skin injury, exhibit limited protrusion, and cover the original site and size of injury with new tissue whose pigmentation is usually lighter than that of the surrounding intact skin. In contrast, keloids are characterized by exuberant, excessive, protruding, and firm new tissue formation above and beyond the site and size of the original skin injury at specific parts of the human body (i.e., the sternum, mandible, and deltoids); the color of the keloids is much darker than that of the surrounding intact skin. Histologically, keloids are characterized by increased rate of collagen synthesis and deposition as well as increased content of glycosaminoglycans and water in the extracellular matrix [5].

Medical treatments of keloids involve topical application of pressure and injections of steroids until there is no excess, abnormal collagen growth at the scar site. Last but not least, keloids can be excised by surgery. Since surgical procedures injure skin, unwanted scars may recur after keloid excision surgery.

3.5.3 Scarless Wound Healing During Early Gestation of the Fetus: A Special Case

A very special, in fact the ideal, case of wound healing occurs only in the fetus during the early stages of gestation. The remarkable outcome (first observed in lambs) is that, during the early stages of gestation, the fetus injured skin healed without scars, an outcome which proved to be age dependent. Propensity for scar formation increased starting as early as the latter stages of gestation and after birth [5, 6]. Similar results from subsequent studies in mice, rats, pigs, monkeys, and humans were also reported in the medical literature [6, 7].

The motivation for research regarding this exceptional natural case of wound healing has been the hope that deciphering the underlying mechanisms could lead to the development of novel clinical treatments in order to obtain scarless wound healing in humans after birth.

3.5.4 Parameters Affecting Scarless Wound Healing

Research conducted to date has identified several parameters pertinent to scarless wound healing. One such parameter is the size of wounds; even in fetal lambs, more than 50% of wounds larger than 4 mm in diameter healed with scars [6].

The characteristics of cellular- and biochemical-level studies reported in the biomedical literature were that (i) specimens from fetal and adult tissues were investigated in parallel, and the respective results were compared, and (ii) the research strategies examined cellular-, molecular-, and gene-level events in each stage of the tissue wound healing process (reviewed in Chapter 2: Table 2.1). Overall, these results provided impressive evidence of several conditions in the fetus which are either different, missing, or contraindicatory during the stages of inflammation, Proliferation/Repair and Remodeling/Maturation of the wound healing process in adults. Identifying the underlying responsible differences may explain the observed adverse results after birth.

In contrast to results obtained from adult tissues, noteworthy differences *in the fetus* include (but are not limited to) the following [6, 8]: (i) mesenchymal stem cells and progenitor cells (which are needed to mediate the scarless tissue repair) are present in higher numbers at the site of injury; (ii) very few inflammatory cells are present during the process of skin wound healing; (iii) Interleukins 6 and 8, bioactive chemical compounds which cause inflammation, are present in low levels in the fetus, but Interleukin 10, which is an anti-inflammatory chemical compound, is present in increased levels; (iv) myofibroblasts are absent from the injury site in fetal tissues; (v) the rate of synthesis of the extracellular matrix components is high; (vi) Collagen Type III, which is predominantly present in the fetus, promotes cell migration and proliferation (essential aspects of new tissue formation); (vi) hyaluronan (expressed at high persistent levels in the fetus) promotes cell migration and traps water; (viii) adhesion proteins (which are fundamental for the survival and function of anchorage-dependent cells responsible for new tissue formation during the wound healing process) are upregulated rapidly and promote cell attachment and migration in the fetus; (ix) among the proteoglycans, decorin and heparan sulfate are either absent or in decreased levels, but chondroitin sulfate is present in increased levels in fetal tissues; (x) collagen in the extracellular matrix of the fetus is deposited in fine, basket-like weave. Collagen cross-linking is low in the fetus; (xi) metalloproteinases (MMPs), which promote accumulation of collagen, turnover of excess extracellular matrix modulators, and tissue remodeling, are high in fetal tissues; (xii) among the Growth Factors (the chemical compounds that promote proliferation as well as other functions of the constituent cells in tissues), Transforming Growth Factor beta-one and Transforming Growth Factor beta-two are present in low levels, but Transforming Growth Factor beta-three is present in high levels. Vascular Endothelial Growth Factor (which promotes initiation and formation of new blood vessels in healing tissues) is present at high levels in the fetus. Hypoxia-inducible Factor (nitric oxide), which is very important for the formation of new blood vessels, is present in high levels; (xiii), and last but not least, genes (which affect cell proliferation) are rapidly upregulated in the fetal tissue [6, 8].

3.5.5 Closing Remarks

Undoubtedly, the aforementioned observations illustrate the major differences between the type and number of cells and of the type, and amount of the extracellular matrix components present in the skin of adults and in the skin of fetuses at the early stages of gestation. Are those aspects some of the secrets responsible for the desirable, ideal outcome of scarless tissue wound healing in the fetus? Further research is needed and continues using the latest biomedical knowledge, methodologies, and technology. Unquestionably, this field has the potential of providing valuable information, which can be used in medical treatments of wounds to attain scarless healing in the tissues of all people after birth.

3.6 Other Undesirable Tissue Wound Healing Outcomes

Undeniably, either too much or too little new tissue formation is neither a desirable nor an ideal outcome of the tissue wound healing process. In fact, a case of pathological, undesirable, exuberant tissue growth is the case of cancer.

By definition, cancer, a disease that affects many organisms, including humans, is characterized by uncontrollable proliferation of cells from various tissues, which become abnormal in form and function and may spread to other parts of the host organism [9].

Carcinogenesis is the process of developing cancer. Any agent, chemical or otherwise, which causes cancer is referred to as a "carcinogen". The predominant characteristics of cancer are "tumors" or "neoplasmas"; these are masses of tissues resulting from abnormal, uncoordinated, and uncontrollable proliferation of cells (via mitosis) and deposition of excessive extracellular matrix [10].

It is important to differentiate between two types of tumors, which are observed in clinical cases, specifically between "benign tumors" (i.e., tissue masses which are not pathological growths) and the "malignant or cancerous tumors". An essential difference that distinguishes the two types of the aforementioned tumors is the morphology of the contained cells: the benign tumors are composed of well-defined cells whose structure is typical of the cells contained in normal, healthy tissues. In contrast, malignant tumors contain cells which have various bizarre structures and morphology and are not typical of the cells present in normal, healthy tissues. The rate of growth is another very important and contrasting characteristic of the two tumor types. The cells in benign tumors grow slowly; they may grow and then stop, or they may regress; most of these cells are not in mitosis. In contrast, cell growth in malignant tumors is erratic: it may be either slow or rapid. These cells exhibit high mitotic activity resulting in high growth rate of abnormal cells. Moreover, benign tumors are well-defined tissue masses with smooth perimeters and distinct forms. In contrast, the perimeter of malignant cancerous tumors is boundless; their cells are invasive and can infiltrate adjacent normal tissues. A distinctive, unique, and exclusive characteristic of the cells from malignant tumors is their capability to migrate away from the original site and form new malignant tumor(s) at other parts of the host body; this process is known as "metastasis". It should be noted that cells of benign tumors do not metastasize [10].

The expectation is that the natural defense mechanisms of the human body (namely the white blood cells and the immune system) recognize the malignant cells as different, dangerous invaders and attack them in order to eliminate them. Malignant cells and tumors, however, have succeeded in evolving in such ways that they elude detection by the human immune defense system, avoid suppression of cell growth, and evade cell death.

Nonmalignant cells in the human body have specific lifespans during which they survive, exist, and function; once they reach that end time point, they die and are replaced by similar, new cells via natural cell proliferation processes. In contrast, malignant cells elude cell death and become "immortal". Immortality is a fundamental and unique characteristic of malignant cells: as a result, cancer cells can replicate indefinitely and survive forever!

In order to survive, every tissue (including malignant tumors) needs the "life-giving" blood supply. In this respect, cancer tumors have developed a remarkable way of establishing new vascular networks to circulate blood inside their tumor masses. Most importantly, and in addition to developing their own internal blood circulation network, malignant tumors have another extraordinary capability to connect their vascular network with the blood circulation of the host. This arrangement assures that malignant tumors survive and grow for the duration of the host lifespan [10].

3.7 Potential Lessons Learned from Malignant Tumor Pathology

In addition to lessons from physiology and biology, pathology provides another venue of information, which biomedical engineers and scientists can gather and incorporate into their endeavors regarding biomedical applications that benefit patients. A pertinent example is vascularization of the malignant tumor mass *and* successful connection of its internal vascular network with the host blood circulation.

Biomedical engineers recognize the crucial importance of vascularization in the formation, survival, and function of new tissue(s) not only during the wound healing process but also in other biomedical applications. For example, tissue engineering constructs must either be, or become, vascularized post implantation into the human body in order to first survive and then successfully grow new tissue and, thus, replace damaged and diseased tissue. Because of its crucial importance, vascularization has been an area of intensive and multifaceted research, but to date, it has had limited success. It is tempting to wonder whether deciphering the secrets of vascularization in malignant tumors could provide the knowledge needed in order to address and obtain vascularization for the long-term success of tissue wound healing processes, tissue engineering applications, etc.

3.8 Cancer and Implants

Biomedical engineers must be concerned about malignant cells and tumors because of the "Delaney Amendment" to the Pure Food and Drug and Cosmetic Act, which was recorded in the Federal Register of the US government in 1977 and is a law in the United States [11]. Part of this amendment states: "...no additive shall be deemed to be safe if it is found to induce cancer when ingested by man or animal or if it is found, after tests which are appropriate for the evaluation of the safety of food additives, to induce cancer in man or animals..." [11].

Because of this law, biomaterials and medical devices, which are implanted in humans and in animals, cannot cause malignant growth of the surrounding tissues or anywhere else in the host body. In consideration for patient safety, as well as because of concerns regarding potential legal liability, any cancer-related indication and outcome ends pertinent research and stop related biomedical product development and production [11].

3.9 The Patient Factor

Last but not least, biomedical engineers (specializing in implantable biomaterials and medical devices) must be aware of the "patient factor".

In order to avoid predictable complications and problems, biomedical engineers should obtain pertinent information from either the scientific biomedical literature or collaborating physicians, surgeons, and health-care practitioners and use it when designing, formulating, and producing the next generation of implant materials and devices.

To date, large-scale productions of profitable, marketable biomedical products (including implantable biomaterials and medical devices) have been designed, formulated, and manufactured based on the model of an "average, young, healthy human being". In fact, the aforementioned products are implanted in patients who are either elderly or in individuals who have various other unrelated pathological conditions. Prescribed medications which these patients receive for other illnesses may either interfere with, or affect, for example, the function of platelets, coagulation factors, white blood cells, etc. For these reasons, biomedical engineers specializing in implants should know not only the physiological but also the pathological milieu in which biomaterials and medical devices will be implanted in order to assure subsequent proper function during the lifespan of the recipient patient(s).

References

1 Goldman L, Cooney KA, editors. *Goldman-Cecil Medicine.* 27th ed. Elsevier; 2023.

2 Schultz GS, Chin GA, Moldawer L, Diegelmann RF. Principles of wound healing. *Diabet. Foot Prob.* [Internet]. 2011 [cited 2023 Sep 1];395–402. Available from: https://www.ncbi.nlm.nih.gov/books/NBK534261/.

3 Thiruvoth F, Mohapatra D, Sivakumar D, Chittoria R, Nandhagopal V. Current concepts in the physiology of adult wound healing. *Plast. Aesthet. Res.* [Internet]. 2015 [cited 2023 Sep 1];2:250–260. Available from: https://parjournal.net/article/view/1211.

4 Ireton JE, Unger JG, Rohrich RJ. The role of wound healing and its everyday application in plastic surgery: a practical perspective and systematic review. *Plast. Reconstr. Surg. Glob. Open* [Internet]. 2013 [cited 2023 Sep 1];1(1). Available from: https://pubmed.ncbi.nlm.nih.gov/25289204/.

5 Rodrigues M, Kosaric N, Bonham CA, Gurtner GC. Wound healing: a cellular perspective. *Physiol. Rev.* [Internet]. 2019 [cited 2023 Sep 1];99(1):665–706. Available from: https://journals.physiology.org/doi/10.1152/physrev.00067.2017.

6 Larson BJ, Longaker MT, Lorenz HP. Scarless fetal wound healing: a basic science review. *Plast. Reconstr. Surg.* [Internet]. 2010 [cited 2023 Sep 2];126(4):1172. Available from: https://www.ncbi.nlm.nih.gov/pmc/articles/PMC4229131/.

7 Hantash BM, Zhao L, Knowles JA, Lorenz HP. Adult and fetal wound healing. *Front. Biosci.* [Internet]. 2008 [cited 2023 Sep 1];13(1):51–61. Available from: https://www.imrpress.com/journal/FBL/13/1/10.2741/2559.

8 Lo DD, Zimmermann AS, Nauta A, Longaker MT, Lorenz HP. Scarless fetal skin wound healing update. *Birth Defects Res. Part C Embryo Today*

Rev. [Internet]. 2012 [cited 2023 Sep 2];96(3):237–47. Available from: https://onlinelibrary.wiley.com/doi/full/10.1002/bdrc.21018.

9 Feitelson MA, Arzumanyan A, Kulathinal RJ, Blain SW, Holcombe RF, Mahajna J, Marino M, Martinez-Chantar ML, Nawroth R, Sanchez-Garcia I, Sharma D, Saxena NK, Singh N, Vlachostergios PJ, Guo S, Honoki K, Fujii H, Georgakilas AG, Bilsland A, Amedei A, Niccolai A, Amin A, Ashraf SS, Boosani CS, Guha GS, Ciriolo MR, Aquilano K, Chen S, Mohammed SI, Azmi AS, Bhakta D, Halicka D, Keith WN, Nowsheen S. Sustained proliferation in cancer: mechanisms and novel therapeutic targets. *Semin. Cancer Biol.* 2015;35(Suppl):S25–S54. Available from: https://doi.org/10.1016/j.semcancer.2015.02.006.

10 Hanahan D, Weinberg RA. Hallmarks of cancer: the next generation. *Cell* [Internet]. 2011 [cited 2023 Sep 2];144(5):646–74. Available from: https://pubmed.ncbi.nlm.nih.gov/21376230/.

11 National Research Council (US) Committee on Diet N AND C. The delaney clause and other regulatory actions. 1982 [cited 2023]; Available from: https://www.ncbi.nlm.nih.gov/books/NBK216642/.

4

Lessons Learned from Biology

For successful biomedical engineering endeavors, biomedical engineers and scientists need a thorough knowledge of the biology and physiology of tissues in order to obtain tissue wound healing and new tissue formation in conjunction with implanted biomaterials and medical devices. For this reason, the biomedical engineers and scientists must use cellular- and molecular-level information, that cells and their immediate external environment (known as the "extracellular matrix"), form a tissue, a biological entity for a specific physiological function, and organs are composed of several tissues [1].

4.1 Mammalian Cells

Cells are the "living" components of all tissues and organs. Examples of *mammalian cells* include endothelial cells (covering the interior surface of the cardiovascular system), osteoblasts (the bone-forming cells), osteoclasts (the bone-resorbing cells), keratinocytes (skin cells), epithelial cells (covering the interior surface of the respiratory and gastrointestinal systems), fibroblasts (ubiquitous cells present in many human tissues), etc. Although each tissue contains several types of cells, only a few are the most representative and the ones responsible for tissue characteristics and function.

4.1.1 Cell Functions

With the exception of blood cells (which, under normal physiological conditions, constantly move in the dynamic blood circulation milieu), all other cells of human tissues are anchorage-dependent. By definition, anchorage-dependent are cells which adhere on substrates (either natural

Concepts of Tissue-Biomaterial Interactions: Fundamentals and New Directions,
First Edition. Rena Bizios and Marissa E. Wechsler.

Companion Website: www.wiley.com/go/Concepts_of_Tissue-Biomaterial_Interaction_1e

or synthetic). In fact, adhesion of anchorage-dependent cells, which is the very first interaction of these cells on all substrates, is necessary for their survival and for their subsequent normal functions, including those pertinent to new tissue formation. Other pertinent functions of cells include adhesion to other types of constituent cells (for tissue cohesiveness), morphology, proliferation, synthesis and deposition of components of the extracellular matrix, differentiation (when applicable as in the case of stem cells), cell-to-cell signaling, apoptosis, etc.

Undoubtedly, the critical need for new tissue formation is live and functional cells. Anchorage-dependent cells have many characteristic functions (i.e., adhesion, migration, proliferation, etc.). First and foremost is adhesion which is critically important for cell survival. Cells adhere on either substrate surfaces or to other neighboring cells through cell-adhesion molecules (i.e., integrins, selectins, cadherins, etc.). Cell morphology differs from tissue to tissue. Under normal, healthy conditions, human cells have specific lifespans and programmed cell death (known as "apoptosis"). Furthermore, cells exhibit various levels of motility or *migration*, as well as *chemotaxis*, a type of directed cell movement in response to chemical stimuli. Most human cells, but not all, have the capability of *proliferation* (which results from cell division), resulting in increased cell numbers. *Differentiation* is a cell function applicable to stem cells, specifically their capability, under appropriate stimuli, to become any other cell in the human body. Last but not least, cells synthesize chemical compounds (proteins, proteoglycans, etc.) which are deposited in, and form, their extracellular matrix.

4.2 The Extracellular Matrix

The *extracellular matrix* of all tissues contains various organic chemical compounds such as proteins (for example, collagen, fibronectin, laminin, vitronectin, etc.), proteoglycans (such as aggrecan, decorin, syndecan-1, etc.), glycosaminoglycans, etc. In bone, the extracellular matrix contains both an organic (mostly Collagen Type I and other non-collagenous proteins) phase and an inorganic phase (consisting of ceramics such as hydroxyapatite).

In all tissues, the extracellular matrix is a multifunctional environment: it supports the constituent cells throughout their lifespan, contains bioactive chemical compounds/stimuli (such as growth factors) which regulate cell functions (such as cell survival, migration, proliferation, and, whenever applicable, differentiation), and provides the mechanical/biophysical milieu that supports and maintains the structure and function of tissues and organs.

Overall, the extracellular matrix of physiological tissues provides the milieu that constituent cells need to survive and function (in order, for example, to form new tissue) [2, 3].

4.3 Brief Introduction to Proteins

With the exception of blood cells, all other human cells are anchorage dependent, meaning they must adhere on a substrate in order to survive and function (e.g., to form new tissue). For this reason, biomedical engineers and scientists need to know the underlying mechanisms regarding how cells interact with materials (natural and synthetic).

Biomedical engineers and biomaterial scientists are especially interested in proteins for the following reasons: (i) All biological liquids of the body contain proteins; for example, blood plasma contains coagulation proteins, complement proteins, and immunoglobulins. In addition, human tissues contain proteins (e.g., collagen, fibronectin, laminin, etc.) in their extracellular matrices; (ii) within seconds following the introduction of biomaterials and medical devices into the human body, they are covered with adsorbed proteins from body liquids and from the extracellular matrices of surrounding tissues; (iii) proteins pre-adsorbed onto substrates mediate adhesion of cells and promote subsequent cell functions (i.e., adhesion, migration, proliferation, synthesis and deposition of organic extracellular matrix chemical components, etc.) pertinent to new tissue formation [4].

The fundamental building blocks of proteins are *amino acids*, organic compounds each containing amine and carboxylic acid functional groups and carbon bonds. In the human body, the 20 naturally occurring amino acids, the "building blocks" of proteins, are alanine, arginine, asparagine, aspartic acid, cysteine, glutamic acid, glutamine, glycine, histidine, isoleucine, leucine, lysine, methionine, phenylalanine, proline, serine, threonine, tryptophan, tyrosine, and valine (Table 4.1).

Amino acids are organic chemical compounds and contain carbon atoms. Carbon has a valence of four and can attach to four other atoms. The simplest structure is that of a single, central carbon atom attached to a carboxyl (COOH) group, an amine (NH_2) group, and a hydrogen atom; the fourth position, designated by R, can accommodate various other atoms specific to each amino acid. Amino acids contain more than one carbon atom in their core linked by peptide bonds formed during pertinent chemical reactions resulting in *polypeptides*. A *polypeptide* is a linear chain composed of many amino acids. *Proteins* are polypeptides which have molecular mass of 10,000 or more Daltons and complex structures [5]. The primary protein structure

Table 4.1 List of amino acid components of the proteins in the human body.

Amino acid subgroup	Amino acid	Three letter abbreviation	One letter abbreviation
Nonpolar	Alanine	Ala	A
	Isoleucine	Ile	I
	Leucine	Leu	L
	Methionine	Met	M
	Phenylalanine	Phe	F
	Proline	Pro	P
	Tryptophan	Trp	W
	Valine	Val	V
Polar	Asparagine	Asn	N
	Cysteine	Cys	C
	Glutamine	Gln	Q
	Glycine	Gly	G
	Serine	Ser	S
	Threonine	Thr	T
	Tyrosine	Tyr	Y
Negatively charged	Aspartic acid	Asp	D
	Glutamic acid	Glu	E
Positively charged	Arginine	Arg	R
	Histidine	His	H
	Lysine	Lys	K

contains a linear array of amino acids, which form loops, the characteristic arrangement of the secondary structure. When the tertiary and quaternary structures of proteins are formed, the respective arrangements of the amino acid loops are more extensive and convoluted. As a result, proteins are big molecules (*macromolecules*) with convoluted structures and high molecular weight [5].

Examples of extracellular matrix proteins which have been studied extensively are collagen (a chemical compound found in many tissues in the human body) and elastin (whose molecules respond to tensile stress by rearranging their random, relaxed orientation into a linear arrangement reflected at the tissue level as stretching) [6]. Fibronectin was the first extracellular matrix protein whose structure revealed the underlying mechanisms of cell interactions with proteins. Elegant, pioneering research cleaved sequential, small sections of the fibronectin (*ligand*) molecule, checked them for adhesion of cultured cells, and revealed bioactive domains (each composed of up to six amino acids). The tripeptide Arginine-Glysine-Aspartic Acid (RGD) was the first peptide discovered, which proved to be the region (*epitope*) recognized by integrin receptors on cell membranes during cell adhesion [7, 8]. The extraordinary importance of RGD was established and expanded when subsequent research provided evidence of the presence of RGD in other adhesive, extracellular matrix proteins, such as collagen, vitronectin, laminin, etc.

It should be noted that other epitopes of fibronectin proved to have affinity for heparin, fibrin, collagen, etc., which are recognized by other receptor types on cell membranes and by chemical compounds important for other body functions. Many other epitopes with select functions have been identified from research of other proteins relevant to various tissues [9, 10].

4.4 Protein Adsorption on Substrates

By definition, adsorption is the process during which molecules of chemical compounds (e.g., proteins) interact with surfaces of substrates (either natural or synthetic). It should be noted that the terms "adhesion" and "adsorption" on material surfaces are used specifically to describe interactions of cells and proteins, respectively.

Several properties of proteins affect their interactions with substrates immersed in protein solutions in water. Since proteins are heavier than water (and body liquids contain mostly water), protein molecules tend to settle, and eventually, some of them come into contact with the material substrate surface. As time progresses, more protein molecules arrive and accumulate on the substrate surface; area limitations there lead to crowded situations, rearrangement of the molecular orientation (to accommodate more molecules interacting with the surface), and eventually, displacing some molecules and propelling them back into the supernatant, bulk solution away from the surface. The protein concentration in bulk solutions affects the number of molecules interacting with the material surface. When only a few molecules are present, they can spread out on the available surface area. In the presence of many adsorbing molecules, however, the crowded situation dictates accommodation and reorientation of protein molecules on substrate surfaces, which may lead to protein conformation changes. Protein unfolding occurs with time as the molecule makes more contacts with the substrate surface. The result is *conformation*, i.e., specific, three-dimensional, structural configuration that proteins can acquire when they adsorb onto substrates.

The process of detachment of protein molecules from surfaces is *desorption*. By definition, this is the process during which protein molecules previously bound onto the surface of a substrate detach and return to the bulk solution. Because desorption requires simultaneously breaking all bonds the adsorbed protein molecule had formed on a surface, it is not easy to achieve; in fact, most of the times, protein molecule desorption is impossible [11].

4.5 Protein Interactions with Substrates: The "Substrate Prospective"

From the substrate perspective, surface properties which affect adsorption of proteins include topography, composition, heterogeneity, and potential. Greater surface texture increases domains for subsequent protein adsorption. Heterogeneity (i.e., nonuniform characteristics) promotes different protein interactions. The chemical composition of material surfaces determines intermolecular forces, which control subsequent protein binding. Hydrophobic surfaces promote increased protein interactions. Surface potential affects ion distribution and subsequent protein interactions [12].

4.6 Adsorption of Proteins from Multicomponent Solutions. The Vroman Effect

Because body liquids (for example, blood plasma) contain several and different proteins, the events which take place when proteins from multicomponent solutions interact with surfaces of materials were also investigated *in vitro* and reported in the pertinent scientific literature.

Proteins from a multicomponent solution (which, by definition, contains more than one type of protein in the same solution) interact with a substrate in a spatiotemporal fashion. In this case, certain proteins (due to their size, concentration, charge, etc.) are the first to arrive on the substrate surface. The time course of arrival of each protein type depends on the size of the molecule and respective protein solution concentration: for example, big molecules in high concentration are favored for early arrival at the substrate surface. Over time, however, other proteins arrive seeking a spot on the same substrate surface. Limitations of surface area trigger competition between the early- and late-arriving proteins, leading to sequential molecule exchange as some of the early-arrived molecules are displaced and pushed back into the supernatant bulk multicomponent solution.

Moreover, each protein has its own time of arrival and time course of interacting (including reaching a maximum) and then desorbing from the surface; eventually, the amount of each adsorbed protein reaches a plateau. At any given time, different types of proteins at various concentrations may be present on the material surface.

Pioneering research in this field was conducted by Leo Vroman, PhD, who studied adsorption of the proteins dissolved in blood plasma, a special multicomponent protein solution (discussed in Chapter 2: Section 2.2.2) on natural and synthetic substrates. Pioneering research by Dr. Vroman elucidated the order of adsorption of select blood plasma proteins (of various molecular weights, sizes, and concentrations) onto synthetic materials [13, 14]. In recognition of Dr. Vroman's notable scientific contributions in this field, his name was given to the phenomenon of interactions of proteins from a multicomponent solution with substrates. The *Vroman Effect* refers to the hierarchical (in space and time) series of collisions of protein molecules, adsorption, and exchange processes when proteins from a supernatant, multicomponent solution interact with the surfaces of material substrates.

4.7 Protein-Mediated Cell Interactions with Substrates

Adhesion of anchorage-dependent cells, the first interaction of these cells on substrates, is necessary for their survival and subsequent normal functions. Successful anchorage-dependent mammalian cell adhesion on material substrates occurs within 1.5–2 hours from the initial contact *in vitro* [15, 16].

Cells do not interact with "bare" material surfaces; they require the mediation of pre-adsorbed proteins in order to adhere to surfaces. The type, distribution, and conformation of these proteins subsequently dictate cell adhesion, survival, and function, ultimately leading to the overall formation and function of new tissue. Implantation of biomaterials and medical devices exposes them to biological liquids and tissues containing many proteins and results in their immediate coverage by adsorbed proteins.

Cells adhere on substrates through their interactions with pre-adsorbed proteins. Receptors on cell membranes recognize and bind to specific domains of extracellular matrix proteins. An example of such receptors is *integrins*, which recognize and bind to various peptide sequences such as the RGD domain of several extracellular matrix proteins, e.g., fibronectin, collagen, etc.

4.8 Protein and Cell Interactions with Materials: Applications

Biomedical scientists and engineers obtained invaluable information from advances in basic science fields such as cell biology, molecular biology, molecular biochemistry, etc., and transferred that knowledge to applications pertinent to implantable materials and medical devices.

Realization that the extracellular matrix of natural tissues provides a unique environment for the survival and function of anchorage-dependent cells (a condition which is not provided by the synthetic biomaterials) became the rationale and justification for research endeavors toward simulating aspects of the natural milieu.

Some pertinent approaches were motivated by the knowledge that (i) proteins pre-absorb on the surfaces of substrate materials and (ii) anchorage-dependent cells subsequently interact with the pre-adsorbed proteins. That information was applied in developing chemical modifications of synthetic materials and aimed at providing the environment for achieving interactions and functions of select, essential cells from the surrounding tissues at the implant site/interface. This requirement is necessary to assure success of implants post implantation.

4.9 Select Chemical Modification of Substrate Surfaces

Since adhesion to substrates is the primary and decisive function for the survival of anchorage-dependent cells, it was the first cell function to be addressed in pertinent studies using *in vitro* models, cultured mammalian cells, and various material substrates. For this purpose, proteins, known to mediate adhesion and subsequent functions of select cells from specific tissues, were chosen and used for the chemical modification of the surface of synthetic materials. The following representative examples illustrate those research endeavors.

4.9.1 Directed Mammalian Cell Adhesion on Chemically-Modified, Protein-Containing, Patterned Surfaces

Electrodes are neural prostheses with pointed insertion edges and surfaces which are not completely covered with an electric circuit. In fact, the electrically active areas are distinct regions connected via circuit lines, both among themselves and to the electric system (located outside the brain). Electrodes (in the order of a few millimeters in length and 1 mm or less in width for experiments with small size animals, e.g., rats) are inserted into the brain (a soft tissue) for one of the following purposes: either (i) to record electric signals inside the brain and transfer signals of electrical activity from the brain to signal-collection monitors outside the body; or (ii) to provide electrical stimulation to specific regions of the brain. Such procedures are conducted either using experimental animals in laboratories for basic research purposes or in hospitals/clinics to stimulate regions of the brains of patients suffering from neurological disorders, hoping to stimulate relevant functions of the brain and, thus, improve the welfare of patients [17].

Neurons, the predominant cells in the brain, transmit electric signals and also respond to electrical stimulation from the outside world. In contrast, the astroglial cells (another type of cells in the brain) do not transmit electric signals and do not respond to electrical stimuli. For this reason, if astroglial cells cover the inserted electrode, transmission of electric signals ends. Similar to physiological responses to implants in other parts of the body, "encapsulation" of the inserted electrode with astroglial cells may be the way the brain "deals" with the "invader" electrode. From the perspective of electrode efficacy, however, such coverage with astroglial cells results in failure of the transmission of electric signals either from or to the brain.

These results became the impetus for researchers in the field of neural electrodes to design, formulate, and test judiciously chosen chemical modifications of material surfaces to promote interactions of neurons and astroglial cells onto different and specific regions of electrode surfaces. In order to achieve directed attachment of cells to predetermined regions, material surfaces were micropatterned with geometric configurations (e.g., squares of various surface areas). Each pattern was modified chemically using N^1[3-(trimethoxysilyl)propyl]diethylenetriamine (DETA), an adhesive hydrophilic organosilane, covering the interior of each pattern. Octadecyltrichlorosilane (OTS), a nonadhesive chemical compound, was deposited on the perimeter of each pattern using established chemical techniques [18]. In subsequent *in vitro* experiments, cultured rat astroglial cells were directed and adhered on the interiors of the patterns modified with the adhesive DETA; moreover, depending on the size of the interior surface area, a single cell was directed to each one of the small (specifically, 50 μm per side) patterns [17, 18]. This outcome with astroglial cells provided important evidence that appropriate chemical modification of substrate surfaces can be used to direct cells for adhesion purposes on prearranged sites on materials.

In parallel but separate experiments, chemical modification of the substrate surface focused on the characteristic function (i.e., electric signal transduction) of neurons. The body of the neurons (known as "soma") extends finger-like projections in a dendritic arrangement, out of which one projection becomes an *axon* (an arm-like extension) through which neurons connect with each other and make contact with distant targets. The resultant linear formation provides the venue required for subsequent transmission of electric signals.

In this case of chemical modification of substrate surfaces, the challenge was to obtain: (i) adhesion of neurons specifically on a pattern simulating an electric circuit (a linear configuration); and (ii) axon outgrowth on prearranged linear tracks of various widths. A critical detail of paramount importance in this study was the knowledge that neurons preferably adhere to polylysine-conjugated laminin [19]. Hippocampal rat neurons adhered to substrates chemically modified with linear tracks of polylysine-conjugated laminin. Most importantly, neuronal axon outgrowth followed faithfully the peregrination of linear tracks as narrow as 2.6 μm in width. These results provided evidence that judiciously chosen chemical modification of substrates promotes cell functions which are pertinent to the predominant cells of a specific tissue (in this case, the neurons of the brain) [19].

4.10 Select Mammalian Cell Adhesion on Chemically-modified, Peptide-Containing, Material Surfaces

In the meantime, advances in the biological sciences (such as cell biology, molecular biology, biochemistry, etc.) elucidated details of the mechanisms involved when mammalian cells interact with proteins (discussed in Chapter 4, Section 4.7) [7, 8]. Pertinent important lessons were that cells do not interact with the entire, big, bulky, convoluted protein molecules. Instead, adhesive epitopes composed of a few amino acids (*peptides*) are recognized by specific receptors on cell membranes; thus, cells interact (e.g., adhere) with proteins pre-adsorbed on substrates [7, 8].

Information regarding the mechanism(s) of cell-protein interaction(s) was applied in a series of *in vitro* experiments using cultured cells and appropriately modified material substrates. Material surfaces were chemically modified using covalently bound, immobilized, select bioactive adhesive peptides (for example, RGD) and established techniques [20] in order to achieve adhesion of specific cells, for example, osteoblasts (the bone-forming cells) and not of fibroblasts (associated with formation of soft tissue).

Cultured rat osteoblast adhesion on immobilized peptides (e.g., RGD; [15]) from adhesive proteins (such as collagen, vitronectin, etc., which are known to mediate osteoblast functions in their physiological milieu) was studied in a series of *in vitro* experiments. It should be noted that 45–50% of osteoblast adhesion involves interaction of integrin receptors on the cell membranes recognizing the RGD peptide in adhesive protein ligands. The other 45–50% of the adhesive mechanism of osteoblasts involves *heparan sulfate proteoglycan* receptors which recognize *heparin-binding domains* of extracellular matrix adhesive chemical compounds [16]. For this reason, Lysine-Arginine-Serine-Arginine (KRSR) (a peptide recognized by the heparan sulfate proteoglycan receptors on cell membranes) was designed, fabricated [21], covalently immobilized on substrate surfaces, and examined in a study comparing adhesion of rat osteoblasts with that of rat fibroblasts. The results of this study provided evidence that both the rat osteoblasts and rat fibroblasts tested adhered on the substrates modified with the immobilized RGD peptide. Fibroblast adhesion on the substrates modified with immobilized KRSR, however, was random, anecdotal, and low [22]. In contrast, the osteoblasts adhered on, and completely covered, the surface area modified with KRSR [22].

4.11 Other Mammalian Cell Functions on Chemically-Modified, Peptide-Containing Material Surfaces

The importance of anchorage-dependent cell adhesion onto substrates (either extracellular matrix proteins or onto implantable biomaterials) cannot be overemphasized: adhesion is critical for the survival of these cells. Once, however, these cells survive, they must also function. Cells have many pertinent functions. During the Proliferation/Repair stage of the tissue wound healing process (discussed in Chapter 2, Section 2.6), for example, the few random cells from the surrounding tissues at the injury site may not be enough to bring about restoration of the injured tissue and formation of new tissue. Many more cells are needed, and that result is achieved by cell proliferation.

A pertinent study was motivated by a need in the cardiovascular field (e.g., vascular grafts), specifically to cover the innermost material surface with normal, healthy, intact endothelium in record time. Such an outcome is *endothelialization* (complete coverage of the interior surface of cardiovascular implants which continuously comes in contact with flowing blood).

This study used cultured bovine pulmonary artery endothelial cells, established techniques to chemically and uniformly modify substrate surfaces with covalently-bound Tyrosine-Isoleucine-Glycine-Serine-Arginine

(YIGSR) (a peptide from laminin, an extracellular matrix protein with which endothelial cells interact), and basic Fibroblast Growth Factor (in the optimal concentration of 1×10^{-9} g/ml). In the absence of soluble growth factor, adhesion of the tested endothelial cells was enhanced on substrates chemically modified with the chosen YIGSR peptide. Coverage of the substrate surface area, however, had many empty spots without adhering cells. In contrast, when the endothelial cells were cultured on the substrate modified with the adhesive peptide from laminin in the presence of basic fibroblast growth factor for the same period of time, the substrate was approximately 95% covered by endothelial cells. Undoubtedly, given a little longer time, proliferation of the adhering cells could have resulted in coverage of the *entire* surface area of the substrate surface with a layer of cells. Endothelization was the desirable outcome of this study that combined modification of the substrate surface (in order to promote adhesion of select cells) and a chemical stimulus (in order to promote cell proliferation) [20]. In this "proof-of-concept" study, the select growth factor was delivered in soluble form, dissolved in the supernatant medium for cultured cells. Alternative modes for the delivery of a chemical stimulus could be achieved by release from either a coating on the substrate surface or from previous incorporation into the bulk of, and released from, synthetic materials.

4.12 Closing Remarks

Success of chemical modification of synthetic material surfaces to promote specific responses from select cells requires thorough knowledge of the cell-related composition in tissues and of details regarding the proteins which mediate constituent cell functions (such as adhesion, migration, proliferation, etc.). As a result of extensive research in many scientific fields, such information is available in the pertinent biomedical, biological, and biochemical literature. Otherwise, researchers who are dealing with a "new" tissue and its constituent cells should conduct appropriate experiments to elucidate the underlying protein–cell mechanism(s) before attempting any endeavor aiming at the chemical modification of implant material surfaces.

Immobilization of adhesive peptides on the surfaces of biomaterials attracted the attention of many biomedical engineers and has resulted in many studies in laboratories across the United States and the world. Many techniques for immobilizing peptides were developed and tested.

Chemical modification of material surfaces was successful when the immobilized, bioactive peptides promoted locally controlled specific cell functions, such as cell adhesion [19, 22, 23], motility [24], and proliferation [20]. The techniques used for chemically modifying material surfaces, however, had certain limitations, including peptide presentation, distribution, concentration, lack of specificity (because various proteins have similar adhesive domains, which are recognized by similar receptors on different cell types from various tissues), reduced, compromised, lost bioactivity in the chemically active and competing milieu of the mammalian body, and sensitivity of proteins and peptides to the conditions of their environment (such as temperature, pH, etc.). Last but not least, chemical modification of substrate surfaces produced synthetic materials which were used in several animal models with limited success. The strong potential and promise of chemically modified materials substrates remain the motivation for continued, advanced research endeavors in this field.

4.13 Protein and Cell Interactions with Nanostructured Materials

Lessons learned from biology provided information regarding various aspects of protein and cell interactions with "natural" substrates. Biomedical engineers and scientists adopted and used that knowledge in modifying chemically (and sometimes topographically) the surfaces of synthetic, implantable biomaterials. Such approaches aimed at making the synthetic material as close as possible to the natural extracellular matrix environment in order to promote subsequent, desirable cell interactions and functions relevant to successful wound healing and new tissue formation. It should be noted that the pertinent chemistry and appropriate techniques used to achieve those modified material outcomes are sophisticated, demanding, and time-consuming. The potential impact on important clinical needs justifies the continuing interest in this field and pertinent research endeavors.

Other advances in biomaterials provided a promising alternative: nanostructured or nanophase materials. Interest in nanostructured materials is substantiated because the organization of the native extracellular matrices of physiological tissues is hierarchical (i.e., macrostructured, microstructured, and nanostructured). For example, bone, a composite material of organic (primarily collagen and other non-collagenous proteins) and inorganic (primarily hydroxyapatite) phases, has a hierarchical structure ranging from the nanoscale to microscale [25]. By definition, nanostructured materials have at least one external dimension or an internal structure (atomic domain or grain size) which is on the scale of 1–100 nm in at least one direction [26, 27]. In contrast, conventional or microstructured materials have grain sizes greater than 100 nm. It is important to note that both the nanostructured and conventional materials have the same chemical composition.

Nanostructured biomaterials have been investigated for applications in implantable prosthetic devices, tissue engineering, tissue regeneration, and several other biomedical fields. When nanostructured biomaterials are implanted inside the bodies of humans and animals, they are expected to function safely and to successfully achieve specific medical objectives for various periods of time (ideally, for the duration of the life of the recipient patient). Such successful outcomes depend on the interactions of implants with the surrounding biological/physiological milieu post implantation.

Biomedical engineers investigated the interactions of proteins of the biological milieu *and* select cells in order to determine the suitability and applicability of nanostructured materials as implants. The first *in vitro* studies in this field investigated adhesion of cultured osteoblasts (the bone-forming cells) and compared results with those obtained from the adhesion of fibroblasts (ubiquitous cells associated with the formation of soft tissue) on ceramics used in bone-related applications, specifically nanostructured hydroxyapatite (a major component of the inorganic phase of bone), as well as on alumina and titania (two ceramics formed by oxidation when the metals aluminum and titanium, respectively, are exposed to an oxygen-containing environment). The cell-related results were compared with those obtained from parallel studies of the aforementioned cells on conventional/microstructured ceramics. Compared to results obtained on respective conventional substrates, osteoblast adhesion on the nanostructured alumina, titania, and hydroxyapatite was enhanced and increased as the size of the nanostructured grain decreased [26]. In contrast, adhesion of fibroblasts decreased on each nanostructured versus respective conventional ceramic tested [27].

This particular result was intriguing because, for the first time, it revealed enhanced, selective adhesion of certain cells (specifically osteoblasts but not fibroblasts) on nanostructured ceramics versus adhesion of those cells on the respective conventional/microstructured ceramics. The excitement regarding nanostructured materials was that, without any special modification, these material formulations promote functions from select mammalian cells which are pertinent to cell survival and other functions relevant to new tissue formation.

Subsequent investigation of the underlying mechanism(s) revealed the critical role pre-adsorbed proteins have on the adhesive interactions of mammalian cells with nanostructured ceramics. Additional justification for studying protein adsorption on biomaterials is the fact that within seconds from exposure of a biomaterial in biological/physiological liquids and tissues (both rich in natural proteins), proteins will absorb onto the biomaterial surfaces. Depending on the type, concentration, distribution, and conformation, the pre-absorbed proteins will subsequently mediate adhesion and other functions of cells from the surrounding tissues. The importance of mediating proteins was validated because, in the absence of adhesive proteins (contained in the required serum in the supernatant cell culture medium), cell adhesion was very low and similar on both the nanostructured and conventional/microstructured ceramic substrates tested for osteoblasts and fibroblasts. In other words, the select adhesion of specific cells on the nanostructured materials was lost [28].

Further investigation of absorption of one protein at a time on nanostructured materials and subsequent interaction of cells with each pre-absorbed protein on alumina surfaces revealed that the mechanism underlying the observed adhesion of select cells was due to the type, concentration, and conformation of the pre-adsorbed protein. Among the proteins tested, absorption of albumin was similar on either conventional or nanostructured alumina, and laminin preferentially adsorbed on conventional alumina. In contrast, collagen, fibronectin, and vitronectin (three proteins known to mediate adhesion of osteoblasts on natural and synthetic substrates) adsorbed preferentially on the nanostructured ceramics tested [27]. This is a remarkable feature of nanostructured materials and a unique advantage because the observed results were obtained by just formulating a biomaterial substrate with the appropriate grain sizes (that is, less than 100 nm in at least one dimension) and no further elaborate chemical, etc., treatment.

Specificity is a unique characteristic of nanostructured materials, which could be an advantage in using these material formulations for biomedical applications. For example, if biomedical engineers can formulate a nanostructured material to specifically adsorb judiciously chosen, select proteins, such an implant can attract and promote interaction of select cells from surrounding tissues *in vivo*, resulting in new tissue formation and success post implantation. Design and formulation of such specialized substrates require knowledge of cell biology and tissue physiology. Preparation of biomaterials still uses methodologies which were developed a long time ago and, since then, have been used in the materials science and engineering fields. If development of novel methodologies is needed in the future, current challenges present opportunities for new contributions in science and engineering. In this respect, nanostructured materials provide alternative, untapped substrates with great potential as implants.

Other studies using polymer–ceramic composites were prompted by the fact that physiologic bone is a natural composite material with an organic (containing collagen and other non-collagenous proteins) and an inorganic (mostly hydroxyapatite) phase with grain sizes in the nanometer range. When cell adhesion was investigated on composites containing poly(L-lactic) acid (a chemical compound which has been used extensively in the biomaterials field) and nanostructured alumina (various weight percent), osteoblast adhesion increased with increased content of nanostructured alumina [29]. In contrast, fibroblast adhesion was low and independent of the nanostructured alumina content in the substrate composite material. Adhesion of both osteoblasts and fibroblasts on composites of poly(L-lactic) acid with varying weight percent of conventional/microstructured alumina was similar and independent of conventional alumina content in the substrate [29]. Moreover, the bending modulus of the composites containing nanostructured alumina increased as a function of nanostructured alumina content to levels comparable to those of human bone [29]. This result was notable because both conventional and nanostructured pure ceramics (which are biocompatible materials) do not exhibit the required mechanical properties for bone implants because they collapse under dynamic loads.

This study was the first to provide important evidence that by (i) promoting desirable interactions of specific cells from surrounding tissues (for example, osteoblasts *but not* fibroblasts) and (ii) by exhibiting improved mechanical properties (comparable to those of the surrounding tissue, e.g., bone), composite materials containing a nanostructured component are novel formulations of great potential and promise for appropriate biomedical applications [29].

4.13.1 Challenges Regarding Implantable Nanostructured Materials

The early excitement regarding the potential of nanostructured materials for biomedical applications has been tempered in the interim because researchers and the general public have concerns regarding the biosafety of nanostructured materials following long-term implantation in humans. Long-term release of particles and molecules from implanted materials and devices in the human body has been of concern regarding potential health and safety issues.

An example is the case of bone, a tissue characterized by the action of two constituent cells: *osteoblasts*, the bone-forming cells, and *osteoclasts*, whose paramount function is resorption of the bone extracellular matrix. In physiology, the normal functions of these two cells maintain bone homeostasis throughout adulthood. The products of bone resorption are excreted through existing, normal elimination processes of the human body.

Osteoclasts maintain their resorption capability when they interact with ceramics *in vitro*. In contrast to results obtained with conventional materials, however, resorption of nanostructured alumina and hydroxyapatite by osteoblasts *in vitro* is greatly enhanced [30]. Such results added to current concerns regarding biosafety of nanostructured biomaterials post implantation.

Concerns regarding long-term stability of nanostructured biomaterials in the physiological milieu post implantation arose because of the type, size, distribution, and fate of nanoparticles (generated from degrading nanostructured biomaterials with grain sizes less than 100 nm). These nanoparticles can not only attach to, but also infiltrate, mammalian cells (which have characteristic dimensions of 10–100 μm). The relevant concerns of medical professionals are related to the type, size, and accumulation of generated nanoparticles locally, systemic distribution (away from the implantation site where these particles originated), interactions with other biochemical reactions *in vivo*, toxicity, tumorigenicity, adverse health effects on the function of vital organs (such as the heart, brain, liver, etc.), and excretion from the human body. Several of the aforementioned unwanted developments can have detrimental health-related effects on organs and tissues at the cellular-, molecular-, and gene-levels. Additional considerations pertain to the safety of excreted and/or accumulated waste of nanostructured biomaterials and nanoparticles in the environment.

4.13.2 Closing Remarks

Nanostructured materials and nanotechnology are developing fields which need to establish appropriate and sound scientific fundamentals, standards, and methodologies for all pertinent aspects: research, manufacturing, and clinical applications. Nanostructured materials and nanotechnology products have great potential and promise, but all issues regarding their safety and efficacy need to be addressed successfully prior to use for biomedical applications. Only when the potential health-related risks are recognized, addressed, and fully resolved, the future of implantable nanostructured biomaterials and nanotechnology applications in medicine will be assured.

References

1 Urry L, Cain M, Wasserman S, Minorsky P, Orr R. *Campbell Biology*. 12th ed. Pearson; 2021.

2 Hall J, Hall M. *Guyton and Hall Textbook of Medical Physiology*. 14th ed. Elsevier; 2021.

3 Nerem RM. Tissue engineering: the hope, the hype, and the future. *Tissue Eng.* 2006;12:1143–50.

4 García AJ, Vega MD, Boettiger D. Modulation of cell proliferation and differentiation through substrate-dependent changes in fibronectin conformation. *Mol. Biol. Cell* [Internet]. 1999 [cited 2023 Sep 2];10(3):785–98. Available from: https://www.molbiolcell.org/doi/10.1091/mbc.10.3.785.

5 Sun PD, Foster CE, Boyington JC. Overview of protein structural and functional folds. *Curr. Protoc. Protein Sci.* [Internet]. 2004 [cited 2023 Sep 2];Chapter 17(1). Available from: https://pubmed.ncbi.nlm.nih.gov/18429251/.

6 Trębacz H, Barzycka A. Mechanical Properties and Functions of Elastin: An Overview. *Biomolecules*. 2023 13(3): 574. Published online 2023 Mar 22. https://www.ncbi.nlm.nih.gov/pmc/articles/PMC10046833.

7 Pierschbacher MD, Ruoslahti E. Cell attachment activity of fibronectin can be duplicated by small synthetic fragments of the molecule. *Nature* [Internet]. 1984 [cited 2023 Sep 2];309(5963):30–3. Available from: https://pubmed.ncbi.nlm.nih.gov/6325925/.

8 Pierschbacher MD, Ruoslahti E. Variants of the cell recognition site of fibronectin that retain attachment-promoting activity. *Proc. Natl. Acad. Sci. U.S.A.* [Internet]. 1984 [cited 2023 Sep 2];81(19):5985–8. Available from: https://pubmed.ncbi.nlm.nih.gov/6237366/.

9 Magnusson MK, Mosher DF. Fibronectin: structure, assembly, and cardiovascular implications. *Arterioscler. Thromb. Vasc. Biol.* [Internet]. 1998 [cited 2023 Sep 2];18(9):1363–70. Available from: https://pubmed.ncbi.nlm.nih.gov/9743223/.

10 Askari JA, Humphries MJ. Cell–matrix interactions, *Encyclopedia of Biological Chemistry*. WJ Lennarz, MD Lane, editors. 2nd ed. Academic Press; 2013, pages 426–429. ISBN 9780123786319. https://doi.org/10.1016/B978-0-12-378630-2.00130-4.

11 Ratner BD, Hoffman AS, Schoen FJ, Lemons JE. *Biomaterials Science: An Introduction to Materials in Medicine*. Elsevier Science; 1997.

12 Zhou HX, Pang X. Electrostatic interactions in protein structure, folding, binding, and condensation. *Chem. Rev.* [Internet]. 2018 [cited 2023

Sep 2];118(4):1691. Available from: https://www.ncbi.nlm.nih.gov/pmc/articles/PMC5831536/.

13 Leonard E, Turitto V, Vroman L. Blood in contact with natural and artificial surfaces. *Ann. N.Y. Acad. Sci.* 1987;516:1–4.

14 Noh H, Vogler EA. Volumetric interpretation of protein adsorption: competition from mixtures and the Vroman effect. *Biomaterials* 2007; 28(3):405–22.

15 Puleo DA, Bizios R. Mechanisms of fibronectin-mediated attachment of osteoblasts to substrates in vitro. *Bone Miner.* [Internet]. 1992 [cited 2023 Sep 2];18(3):215–26. Available from: https://pubmed.ncbi.nlm.nih.gov/1392695/.

16 Puleo D. *Interaction of Osteoblasts with Orthopedic Biomaterials In Vitro.* Rensselaer Polytechnic Institute; 1991.

17 Kam L. *Modulation of Neuron and Astroglial Cell Function by Micropatterning and Immobilization of Select Biomolecules on Biomaterial Surfaces.* Rensselaer Polytechnic Institute; 1999.

18 Kam L, Shain W, Turner JN, Bizios R. Correlation of astroglial cell function on micro-patterned surfaces with specific geometric parameters. *Biomaterials* [Internet]. 1999 [cited 2023 Sep 2];20(23–24):2343–50. Available from: https://pubmed.ncbi.nlm.nih.gov/10614940/.

19 Kam L, Shain W, Turner JN, Bizios R. Axonal outgrowth of hippocampal neurons on micro-scale networks of polylysine-conjugated laminin. *Biomaterials* [Internet]. 2001 [cited 2023 Sep 2];22(10):1049–54. Available from: https://pubmed.ncbi.nlm.nih.gov/11352086/.

20 Dee KC, Andersen TT, Bizios R. Enhanced endothelialization of substrates modified with immobilized bioactive peptides. *Tissue Eng.* [Internet]. 1995 [cited 2023 Sep 2];1(2):135–45. Available from: https://pubmed.ncbi.nlm.nih.gov/19877922/.

21 Dee KC, Anderson TT, Bizios R. Design and function of novel osteoblast-adhesive peptides for chemical modification of biomaterials. *J. Biomed. Mater. Res.* 1998;40(3):371–7.

22 Hasenbein ME, Andersen TT, Bizios R. Micropatterned surfaces modified with select peptides promote exclusive interactions with osteoblasts. *Biomaterials* [Internet]. 2002 [cited 2023 Sep 2];23(19):3937–42. Available from: https://pubmed.ncbi.nlm.nih.gov/12162326/.

23 Kam L, Shain W, Turner JN, Bizios R. Selective adhesion of astrocytes to surfaces modified with immobilized peptides. *Biomaterials* [Internet]. 2002 [cited 2023 Sep 2];23(2):511–5. Available from: https://pubmed.ncbi.nlm.nih.gov/11761172/.

24 Olbrich KC, Andersen TT, Blumenstock FA, Bizios R. Surfaces modified with covalently-immobilized adhesive peptides affect fibroblast population motility. *Biomaterials* [Internet]. 1996 [cited 2023 Sep 2]; 17(8):759–64. Available from: https://pubmed.ncbi.nlm.nih.gov/8730959/.

25 Gong T, Xie J, Liao J, Zhang T, Lin S, Lin Y. Nanomaterials and bone regeneration. *Bone Res.* 2015 [cited 2023 Aug 12];3(1):1–7. Available from: https://www.nature.com/articles/boneres201529.

26 Webster TJ, Ergun C, Doremus RH, Siegel RW, Bizios R. Enhanced functions of osteoblasts on nanophase ceramics. *Biomaterials* 2000; 21(17):1803–10.

27 Webster TJ, Ergun C, Doremus RH, Siegel RW, Bizios R. Specific proteins mediate enhanced osteoblast adhesion on nanophase ceramics. *J. Biomed. Mater. Res.* 2000 [cited 2023 Aug 12]; 51(3):475–483. Available from: https://pubmed.ncbi.nlm.nih.gov/10880091/.

28 Webster TJ. *Design, Synthesis, and Evaluation of Nanophase Ceramics for Orthopaedic/dental Applications.* Rensselaer Polytechnic Institute; 2000.

29 McManus AJ, Doremus RH, Siegel RW, Bizios R. Evaluation of cytocompatibility and bending modulus of nanoceramic/polymer composites. *J. Biomed. Mater. Res. Part A* [Internet]. 2005 [cited 2023 Aug 12]; 72A(1):98–106. Available from: https://onlinelibrary.wiley.com/doi/full/10.1002/jbm.a.30204.

30 Webster TJ, Ergun C, Doremus RH, Siegel RW, Bizios R. Enhanced osteoclast-like cell functions on nanophase ceramics. *Biomaterials* 2001;22(11):1327–33.

5

Tissue Wound Healing in the Presence of Implants

5.1 Introduction

Injury to tissues causes wounds characterized by cell death, denaturation/destruction of the extracellular matrix, loss of blood vessel integrity, and damage to nerves. Overall, tissue injury breaks the continuity of tissues (Figure 2.1).

During the process of tissue wound healing in normal physiology, what nature has working to its advantage is the fact that on both sides of the injury-induced discontinuity, there are "similarities", – specifically the proteins, cells, biological liquids, blood vessels, and nerves (Figure 2.2). For this reason, although it is still a formidable task, the process of "normal" tissue wound healing is assisted and expedited, often resulting in the best outcome. Successful tissue wound healing involves cell regeneration, synthesis, and deposition of new extracellular matrix, formation of new blood vessels, and restoration of nerves (Figure 2.1). Combined, these repairing events heal the injury by resulting in the formation of new tissue.

Introduction to biomedical devices composed of implantable materials into the bodies of humans and animals requires one or more surgical procedures, which are necessary and unavoidable, *but* inevitably cause cell, tissue, and organ injury. In the presence of implantable materials and/or biomedical devices, the situation at the site of tissue injury is completely different (Figure 2.3). On one side of the tissue, the injury-induced discontinuity, there is still the biological physiological environment with cells, extracellular matrix, biological liquids, blood vessels, and nerves. On the other side, however, there are biomedical devices composed of synthetic materials (i.e., ceramics, metals, polymers, and respective composites). In such cases, therefore, the tissue injury site separates the biological environment from the synthetic material/device and forms a *tissue-implant interface.*

Concepts of Tissue-Biomaterial Interactions: Fundamentals and New Directions,
First Edition. Rena Bizios and Marissa E. Wechsler.

Companion Website: www.wiley.com/go/Concepts_of_Tissue-Biomaterial_Interaction_1e

The most notable difference across the interface between the physiological/biological environment and the synthetic material device is the middle part; specifically the extracellular matrix at the biological site. It is replaced by biomaterials, such as ceramics, metals, polymers, and/or respective composites at the implant site (Figure 2.3). This is the major difference between the milieu of tissue wound healing in the normal biological/physiological environment and in the presence of implantable biomaterials and medical devices. The known outcomes at each stage of the tissue wound healing process in the absence of implants were discussed in Chapter 2 (Figure 2.1); they will now be discussed, compared, and contrasted with the known outcomes of the main stages of the wound healing process (Table 2.1) in the presence of implants.

5.2 Blood Coagulation and Fibrinolysis in the Presence of Implants

Because the human body is extensively vascularized, injuries (including the ones associated with surgery implantation procedures) cause damage to the blood vessels which result in blood leakage outside the vasculature and formation of blood clots (thrombi). Vasculature tissue damage and blood clot formation occur at the tissues surrounding implanted biomaterials and medical devices. In physiology, these events are successfully addressed during the subsequent stages of the process of normal tissue wound healing [1].

Blood clotting is the major and critical complication in the case of cardiovascular implants because blood must circulate in liquid form while it is in constant contact with the synthetic materials used for cardiovascular implants. Although blood clotting is pertinent, and of paramount importance to all cardiovascular implants, the comments in this Chapter will focus on vascular grafts. These are synthetic, tubular structures (usually composed of polymers) which are used clinically to replace either blocked blood vessels (because of blood clots) or blood vessels, which are injured, collapsed, or afflicted with various pathological conditions. The pertinent clinical pathological conditions may require surgery to remove the diseased blood vessel and replace it with a synthetic implant (vascular graft) to reinstate the disrupted, blocked blood flow.

Of paramount importance and priority in the case of cardiovascular implants is *hemocompatibility*, the blood compatibility of the synthetic biomaterials used for cardiovascular applications. In normal physiology, such a desirable outcome is achieved only when blood circulating within the cardiovascular system comes in contact with normal, healthy, intact endothelium (the cell lining covering the innermost surface of the entire cardiovascular system) [2, 3].

The major limitation of current biomaterials used in cardiovascular applications, therefore, is that they do not have the properties of normal, healthy, intact endothelium. Inevitably, when whole blood inside a vascular graft comes in contact with the synthetic materials of cardiovascular implants, the following events take place: first, platelets coming into contact with the synthetic material become activated and form the platelet plug. Simultaneously, the plasma coagulation cascade is activated, resulting in the formation of a fibrin network, which consolidates and stabilizes the platelet plug. The combination of the aforementioned two events results in the formation of a blood clot (thrombus). In addition, activation of other chemical compounds, such as the complement proteins (which are dissolved in blood plasma), results in white blood cell activation.

Remarkable aspects of normal physiology are the checks and balances, which are prerequisites for, and contributors to, hemostasis. One of those aspects is prevention of uncontrolled blood clotting (reviewed in Chapter 3, Section 3.2). This outcome is achieved with the help of naturally occurring anticoagulant chemical compounds, specifically heparin and antithrombin III. The bioactivity of the heparin/antithrombin III complex is a remarkably fast and effective process that inactivates various steps of the plasma coagulation cascade, thus, preventing consolidation of blood clots. Unavailability of thrombin prevents activation of platelets and, thus, formation of the platelet plug (reviewed in Chapter 2, Section 2.3.3: Figure 2.6). Once blood clots are formed, however, antithrombin III and the heparin/antithrombin III complex cannot affect them. The process which nature has to remove and eliminate blood clots (once the purpose of their existence has been successfully met) is *fibrinolysis* (reviewed in Chapter 2). This process sequentially breaks down the fibrin mesh which consolidates and stabilizes the platelet plug (reviewed in Chapter 2); finally, the degradation products are small enough to be removed from the body via appropriate, naturally occurring excretion processes [4].

In contrast, the natural checks and balances are absent when whole blood comes into contact with synthetic materials (for example, inside a vascular graft). The synthetic materials promote blood clot formation. Moreover, the synthetic materials do not have fibrinolytic potential. For these reasons, once blood clots are formed in conjunction with synthetic materials, they remain *in situ* (and may even grow).

5.3 Inflammation in the Presence of Implants

The presence of implants instigates major responses and events, which are completely different from those observed during the inflammation stage of the normal tissue wound healing process in the absence of implants. During these events, white blood cells (leukocytes) are the "protagonists". The white blood cells defend the human body from invading pathogens (reviewed in Chapter 2), but they also defend the human body against any other intruders, including implanted biomaterials and medical devices. In contrast to the processes of coagulation and fibrinolysis, which take place inside, or in conjunction with, the vasculature and the cardiovascular system, the process of inflammation may occur outside the vasculature and inside tissues and organs. For this purpose, the activated neutrophils and monocytes leave the circulating blood, transverse the vascular wall tissue, and migrate to the site of implantation in a manner similar to that used in the response of leukocytes to invading pathogens (described in Chapter 2, Section 2.5.5: Figure 2.9A).

The role of monocytes/macrophages is of paramount importance during inflammation in the presence of implants. Current knowledge of the function of macrophages in those cases started with the pioneering research of Professor James M. Anderson, MD/PhD, and his research team. Their work elucidated the events which take place when macrophages arrive at the site of the implanted biomaterials and medical devices. Pertinent publications by this research team are "classic" communications in the literature of this scientific/medical field [5, 6].

Macrophages arrive at the site of implantation driven by chemotactic responses along concentration gradients of the chemical compounds released from tissues damaged during implantation surgery procedures. By that time, the implanted biomaterials/medical devices are exposed to the physiological milieu rich in proteins (from body liquids and surrounding tissues) and are covered by adsorbed proteins [5, 6].

Once the macrophages arrive at the implantation site, they are activated, interact with the pre-adsorbed proteins, adhere, and spread on the implant surfaces (Figure 2.9B). Adherent macrophages fuse and form *foreign body giant cells (Figure 2.9C)*. Compared to the size of either a single monocyte or a single macrophage, foreign body giant cells are big, multinucleated cells; these hallmarks prove that many macrophages have fused to become one, big, foreign body giant cell (described in Chapter 2: Figure 2.9). The pioneering work of Professor James A. Anderson and his research team provided evidence that foreign body giant cells are formed during the inflammation stage of the tissue wound healing process in the presence of implants. Moreover, macrophages are the most important participants in, and contributors to, *the foreign body reaction* – a response that has been documented for all implanted materials and medical devices in the human body [5, 6].

The ultimate function of macrophage responses to invading pathogens is phagocytosis, which is achieved by engulfing the invading microorganisms, bombarding them with several chemical compounds (such as oxygen radicals, hydrogen peroxide, acids, lysosomes, etc.), and killing them. Such elimination of the invading pathogens (reviewed in Chapter 2) successfully defends the human body. What is critical for this desirable outcome is the relative size of the cells involved in phagocytosis: a leukocyte, with characteristic dimensions on the order of 100 μm, has the capability of completely engulfing and incorporating into its cytoplasm a bacterium and a virus, whose characteristic dimensions are on the order of 10 μm and 100 nm, respectively. The subsequent "bombardment" with bioactive chemical compounds and killing of the invading pathogens take place completely inside the enclosed cytoplasmic environment of macrophages.

The situation is very different when macrophages encounter implantable materials and medical devices: they become activated, attempt to engulf the foreign invader, and subsequently, during phagocytosis, destroy/eliminate it with the armamentarium of bioactive chemical compounds they have from nature. A single macrophage, on the order of 21 μm [7], and even the foreign body giant cells (formed in a desperate attempt to increase their cell size in order to engulf implants) cannot succeed in completing the engulfment process because the implant size is in the order of at least several or more centimeters. Engulfment of the implant is thus incomplete, *but* the release of the bioactive chemical compounds by the activated foreign body giant cells proceeds as programmed by nature. This process, known as *frustrated phagocytosis*, is characterized by unsuccessful elimination of the invader implant. From the physiological/biological perspective, all bioactive chemical compounds (i.e., enzymes, oxygen radicals, hydrogen peroxide, acids, lysosomes, etc.), which are produced by the foreign body giant cells during frustrated phagocytosis, are not restricted inside the cell cytoplasm; in fact, they are released, and accumulate, into the surrounding tissues where they can cause local inflammation and tissue injury (including death of the constituent cells). From the material perspective, the aforementioned release of bioactive chemical compounds may interact with the chemical composition of the implanted materials and cause release of ions from metal implants and of monomers from polymers. The released bioactive chemical compounds may also alter the topography of the material surfaces (by forming cracks, pits, etc.). Furthermore, the combination of chemical and topographical effects may cause loss of the integrity, shape, and properties of implanted biomaterials and medical devices, as well as impairment and/or loss of the function of medical devices. In addition to the released bioactive chemicals during frustrated phagocytosis, particles released from the compromised implanted biomaterials may cause local inflammation. If some of these chemicals or particles are transported to other parts of the body, inflammation may spread systemically. Last but not least, activation of leukocytes (such as neutrophils, monocytes/macrophages, and lymphocytes) during chronic inflammation may result in rejection of implants by the human body [5, 6].

In summary, implantation of biomaterials and medical devices used for medical treatments of injury and disease is associated with inevitable tissue injury due to the required implantation surgery procedure(s). The various stages of the "normal" tissue wound healing process in the absence of implants (as reviewed in Chapter 2) are affected by the presence of implanted biomaterials and medical devices. During the inflammation stage, the monocytes/macrophages are the important participating leukocytes. The foreign body reaction is a characteristic, unavoidable result around implants. In this case, the macrophages are the "protagonist" cells; they form the foreign body giant cells at the material/tissue interface in a desperate attempt to increase their size and succeed in engulfing the implant. The subsequent frustrated phagocytosis is the result of the unsuccessful outcome of the natural phagocytosis process; release and accumulation of bioactive chemical compounds in the local surroundings affect survival of tissues and constituent cells, as well as the chemistry, integrity, and function of implanted biomaterials and medical devices.

5.4 Proliferation/Repair and Remodeling/Maturation in the Presence of Implants

The Proliferation/Repair and Remodeling/Maturation stages of the tissue wound healing process in the presence of implants may result in undesirable outcomes of clinical importance. Examples of such outcomes are briefly reviewed in the sections that follow.

5.4.1 Chronic Nonhealing Wounds

Chronic, nonhealing wounds are clinical complications associated with pathological conditions such as advanced stages of diabetes, pressure ulcers (e.g., bedsores) on the back of disabled/paralyzed patients (due to their prolonged immobilization in one, often lying, position in bed), etc. Such conditions, characterized by chronically open and nonhealing wounds, result from the combination of detrimental effects on cells and their extracellular matrix, the fundamental constituents of surrounding tissues. Under these conditions, cells either may be present in small numbers or may not be fully functional. Another critical aspect pertinent to cell function in these clinical conditions is limited synthesis and deposition of a functional extracellular matrix. Chronic, nonhealing wounds around implants are undesirable outcomes because they affect the welfare of recipient patients and the fate of implants [8].

5.4.2 Clinical Complications Regarding Vascular Grafts

Following surgical excision of a diseased or clotted segment of a blood vessel, implantation of a synthetic vascular graft substitute requires suturing the two ends of the graft with two ends of the native vascular tissue to form a continuous conduit for blood flow. For a successful outcome in such cases, the synthetic biomaterial must both come into close contact and also mesh with the native vascular tissue; these connecting locations are known as *anastomotic sites*. The structure, dimensions, and composition of the vascular wall tissue were discussed in Chapter 2, Section 2.5.5 and are illustrated in Figure 5.1.

A pathological condition, which is associated with an undesirable outcome to vascular grafts, is *hyperplasia*. Hyperplasia is characterized by increases in the size of organs and/or tissues due to (i) increased proliferation of non-pathological constituent cells and (ii) increased deposition

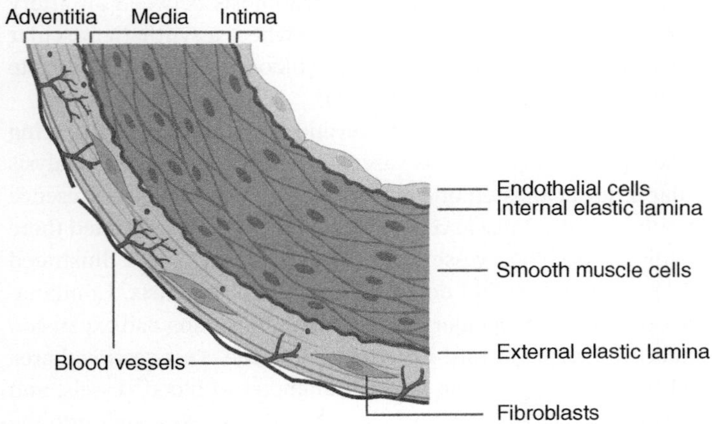

Figure 5.1 Schematic illustration (not to scale) of a cross-section of the arterial wall tissue structure and composition.

of an extracellular matrix by these cells. In the case of vascular tissue, hyperplasia (referred to as "intimal hyperplasia") is a result of the proliferation and migration of vascular smooth muscle cells (which are present in media), causing increased thickening of the vascular tissue inward into the lumen and, thus, a decreased cross-sectional area for blood flow. Intimal hyperplasia, reported in major cases of vascular pathological conditions, is also associated with interventions following either stent implantation or angioplasty procedures; these procedures involve insertion of probes and catheters (composed of synthetic materials) into blood vessels during clinical diagnostic and/or medical treatment procedures [9].

In addition to pathological complications, neointimal hyperplasia has been associated with cases of great interest to the biomedical community relevant to biomedical engineering applications. Mechanical injury of arteries due to stretching of the arterial vascular tissue (for example, when balloon catheters are used in clinical procedures) is associated with neointimal hyperplasia resulting from damage to the endothelial lining and subsequent recruitment of monocytes/macrophages and neutrophils to the site of the injured vascular wall tissue. Activated macrophages in the media release growth factors, cytokines, and other chemical compounds, which stimulate cell migration and proliferation into the intima [10].

Another crucial aspect of the vascular graft implantation *in vivo* is the events taking place at the anastomotic sites. In medicine and surgery, *anastomosis* refers to the joining or communication between two tubular organs. In addition, anastomosis refers to the formation of surgical, traumatic, and pathological formations at the connection between two normally distinct structures, for example, anastomosis between an artery and a vein. Anastomosis also describes the site where a synthetic vascular graft is connected (via surgical sutures) to the blood vessel tissue. The site of that connection is the *anastomotic site* [11].

Anastomotic intimal hyperplasia is of relevance to biomedical engineering cardiovascular applications (such as vascular grafts). Histological analysis of the vascular tissue at anastomotic hyperplasia sites revealed the presence of many vascular smooth muscle cells, which had obviously migrated there from the media of the blood vessel wall tissue (Chapter 2, and illustrated in Figure 5.1), proliferated, and deposited extracellular matrix. Combination of these cellular and molecular events causes thickening and expansion of the vascular wall tissue inward into the lumen (the cross-sectional area where blood flows), decreases the internal diameter of blood vessels, and thus obstructs normal blood flow and blood circulation. Research into the etiology of anastomotic hyperplasia provided evidence that the mismatch between the mechanical properties of the synthetic implanted vascular graft material and the physiologic vascular tissue is exacerbated in the dynamic milieu of blood flow and is responsible for the observed hyperplasia results at anastomotic sites [9, 11].

5.4.3 Closing Remarks

The presence of implant biomaterials and medical devices affects the process of wound healing, which is different from the one observed in the absence of implants (reviewed in Chapter 2). Following implantation of biomaterials and medical devices and the unavoidable injury (due to the required surgical procedures), the tissue wound healing process involves blood–material interactions, formation of a provisional matrix, inflammation (acute and chronic), foreign body reaction (and subsequent frustrated phagocytosis), and fibrous encapsulation [12]. Formation of a fibrous capsule around implants is another inevitable outcome, which has been recorded for implants excised from the human body (after either implant failure *in vivo* or death of the recipient patient [12]).

5.5 Effects of the Biological/Physiological Milieu on Biomaterials

Unquestionably, the presence of implant biomaterials and medical devices *in vivo* elicits responses from the surrounding tissues that may affect the progress and, eventually, the success of implantation procedures and the fate of implants. At the same time, the biological/physiological milieu [13] (illustrated by the mechanical environment of endothelium in the human blood circulation, Figure 5.2) may affect several characteristics and properties of implants.

Synthetic biomaterials implanted in the human body are exposed to physiological pH and temperature, as well as to an environment which is rich in bioactive chemical stimuli (such as enzymes, oxygen radicals, and other compounds) released from surrounding injured tissues and produced during the various stages of the tissue wound healing process. Such bioactive chemical compounds may affect the chemical composition and texture of the surfaces of implanted synthetic materials; for example, chemical molecules, such as corrosion products from metals and monomers leaching out from polymers, may accumulate both in the interface and between the implant and the surrounding tissues. In such cases, they may react on the surface of implants, altering the chemical composition, structure, and texture of implanted biomaterials, affecting the mechanical material properties and interfering with the mechanical and electrical function of implanted devices.

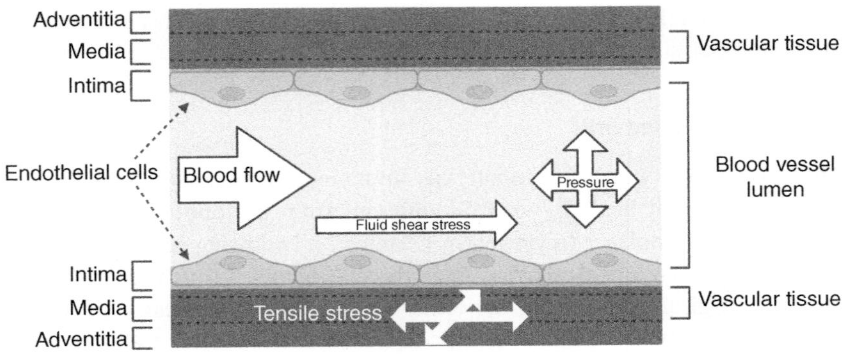

Figure 5.2 Schematic illustration (not to scale) of the mechanical forces in the vasculature. Source: Adapted from Shin [13].

5.6 The Fate of Implants

In addition to affecting the various stages of the tissue wound healing process, the presence of implants *in vivo* may also affect the overall outcome of pertinent medical/clinical treatments. Such responses cover a spectrum of possibilities, including desirable but also undesirable and abnormal outcomes ranging from an acute response (sometimes undesirable) to desirable adaptation (when the surrounding tissues remodel and adapt to reflect changes in demand and function). Abnormal outcomes are not desirable but occur and cause clinical complications.

The most desirable outcome is a resolution, that is, a final state where there are no more biological/physiological changes; this is a "steady state" situation where no changes occur as a function of time.

Examination of implants excised from various parts of the human body revealed a "capsule" of tissue completely surrounding implants composed of synthetic materials. This capsule is known as *fibrous capsule*, and the process leading to its formation as "fibrosis". The fibrous capsule isolates the implant (and the foreign body reaction it elicits) from the local milieu. Encapsulation may be viewed as a solution used by the recipient body to deal with the "implant invader", which cannot be killed and eliminated under other physiological processes (such as phagocytosis by leukocytes). The fibrous capsule covers the implant with tissue the host body recognizes as "its own" and, thus, affords to ignore the implant under the fibrous camouflage.

Biomedical engineers specializing in implantable biomaterials and medical devices have investigated various aspects regarding formation of fibrotic capsules post implantation and identified several pertinent parameters. In this respect, local aspects include the extent of the original injury, the anatomical location (i.e., the organ and tissue site of implantation), the extent of damage (especially destruction of key constituent cells of the surrounding tissues), adequate supply of blood (therefore, adequate supply of oxygen and nutrients to the healing tissue post implantation), the age of the patient, and infection.

Aspects which affect the thickness of the fibrotic capsule include (i) the rate of release and concentration of small molecules such as corrosion ions from metals and monomers leached from polymers; (ii) the detrimental effects of the released chemicals, some of which may be cytotoxic (that is, they kill cells), and others may not kill cells *but* may affect functions of cells in the surrounding tissues needed for tissue wound healing and new tissue formation; (iii) mechanical forces arising, for example, from the relative motion between an implant and the surrounding tissues; (iv) implant design details (such as shape), which affect the fibrotic capsule characteristics (for example, thicker fibrotic capsule over edges and over sharp changes of implants [14]); and (v) electric currents associated with heart and neural devices/applications (such as stimulating electrodes) that promote thickening of the fibrotic capsules around those implants [14].

In addition to local parameters, systemic aspects are also contributors to the formation of fibrotic capsules. These aspects include, but are not limited to, the following: genetics, nutrition, preexisting diseases (specifically chronic diseases such as diabetes and alcoholism), other health-affecting conditions (such as stress, smoking, obesity, etc.), various medications (taken by patients for unrelated conditions but may be either favorable or detrimental toward the formation of fibrotic capsules), and infection.

A special case related to the fate of some implants is *extrusion*. If an implant is located under or very close to skin after the fibrotic capsule is formed and as a result of the relative movement of the surrounding muscles and motion of the skin, implant extrusion is a possible outcome. During this process, the body may slowly push that implant outward through the skin and, eventually, outside of the recipient body [15].

A special case of great interest and importance in the field of tissue engineering is desirable *resorption* of the synthetic material substrates (known as "scaffolds") during the process of wound healing and formation of new tissue. Tissue engineers design, synthesize, and use materials, which (following implantation) will degrade by design and will be resorbed in the physiological environment of the body in synchrony with new tissue formation. The degrading synthetic material products are designed to be excreted via normal physiological excretion processes. It should be emphasized that the aforementioned requirements for tissue engineering applications are specific for the needs of that field. Degradation and resorption are designed, and incorporated, in the synthetic material scaffolds; in other words, such outcomes are not left either to a random or to an unpredictable outcome after implantation. Most importantly, for successful outcomes, it is imperative that the synthetic material scaffold degradation and resorption take place *in synchrony* with the formation of new tissue as the wound healing process progresses and is completed. Early disappearance of a substrate/scaffold before new tissue formation could be disastrous because anchorage-dependent cells, which (with the exception of blood cells) are the cells from all other tissues and organs in the body, *need a substrate to adhere in order first to survive and subsequently to function*. Delayed scaffold material degradation interferes and even prevents new tissue formation, the foremost desirable outcome of tissue engineering endeavors.

The need for novel synthetic biomaterials, which (after implantation in the body) degrade and are resorbed *in synchrony* with new tissue formation, has not been resolved yet. This aspect is still a challenge, but it is also an opportunity for creative ideas, new developments, and contributions by biomedical engineers and scientists in the future. It should be emphasized that, with the exception of tissue engineering applications, resorption of biomaterial substrates post implantation is neither required nor desirable for other medical devices used for tissue engineering applications.

References

1 Williams Hematology, 10th ed. | AccessMedicine | McGraw Hill Medical [Internet]. [cited 2022 Mar 16]. Available from: https://accessmedicine-mhmedical-com.libproxy.uthscsa.edu/book.aspx?bookid=2962.
2 Van Hinsbergh VWM. Endothelium—role in regulation of coagulation and inflammation. *Semin. Immunopathol.* [Internet]. 2012 [cited 2022 Oct 31];34(1):93. Available from: https://www.ncbi.nlm.nih.gov/pmc/articles/PMC3233666/.
3 Yau JW, Teoh H, Verma S. Endothelial cell control of thrombosis. *BMC Cardiovasc. Disord.* [Internet]. 2015 [cited 2022 Oct 31];15(1). Available from: https://www.ncbi.nlm.nih.gov/pmc/articles/PMC4617895/.
4 Chapin JC, Hajjar KA. Fibrinolysis and the control of blood coagulation. *Blood Rev.* [Internet]. 2015 [cited 2022 Oct 31];29(1):17. Available from: https://www.ncbi.nlm.nih.gov/pmc/articles/PMC4314363/.
5 Anderson JM. Mechanisms of inflammation and infection with implanted devices. *Cardiovasc. Pathol.* 1993;2(3):33S-41S.

6 Anderson JM, Rodriguez A, Chang DT. Foreign body reaction to biomaterials. *Semin. Immunol.* 2008;20(2):86–100.

7 Krombach F, Münzing S, Allmeling AM, Gerlach JT, Behr J, Dörger M. Cell size of alveolar macrophages: an interspecies comparison. *Environ. Health Perspect.* [Internet]. 1997 [cited 2023 Aug 29];105(Suppl 5):1261–3. Available from: https://pubmed.ncbi.nlm.nih.gov/9400735/.

8 Rodrigues M, Kosaric N, Bonham CA, Gurtner GC. Wound healing: a cellular perspective. *Physiol. Rev.* [Internet]. 2019 [cited 2022 Oct 31];99(1):665–706. Available from: https://journals.physiology.org/doi/10.1152/physrev.00067.2017.

9 Subbotin VM. Analysis of arterial intimal hyperplasia: review and hypothesis. *Theor. Biol. Med. Model.* [Internet]. 2007 [cited 2022 Oct 31];4. Available from: https://pubmed.ncbi.nlm.nih.gov/17974015/.

10 Lee S, Lee RT. Editorial: Mechanical stretch and intimal hyperplasia: the missing link? *Arterioscler. Thromb. Vasc. Biol.* [Internet]. 2010 [cited 2022 Oct 31];30(3):459–60. Available from: https://www.ahajournals.org/doi/abs/10.1161/ATVBAHA.109.201509.

11 Bassiouny HS, White S, Glagov S, Choi E, Giddens DP, Zarins CK. Anastomotic intimal hyperplasia: mechanical injury or flow induced. *J. Vasc. Surg.* [Internet]. 1992 [cited 2022 Oct 31];15(4):708–17. Available from: http://www.jvascsurg.org/article/0741521492900195/fulltext.

12 Anderson JM. Inflammation, wound healing, and the foreign-body response. B. D. Ratner, A. S. Hoffman, F. J. Schoen, J. E. Lemons, editors. *Biomaterials Science An Introduction to Materials in Medicine.* 3rd ed. 2013;503–12.

13 Shin HY. *Cyclic Pressure Modulates the Endothelial Cell Phenotype: Examination of Pressure-sensitive Mechanotransduction at the Cellular, Transcriptional, and Molecular Levels.* Rensselaer Polytechnic Institute; 2003.

14 Carnicer-Lombarte A, Chen ST, Malliaras GG, Barone DG. Foreign body reaction to implanted biomaterials and its impact in nerve neuroprosthetics. *Front. Bioeng. Biotechnol.* 2021;9:622524.

15 Smith GA, Pace J, Corriveau M, Lee S, Mroz TE, Nassr A, et al. Incidence and outcomes of acute implant extrusion following anterior cervical spine surgery. *Glob. Spine J.* [Internet]. 2017 [cited 2023 Aug 29];7(1 Suppl):40S. Available from: https://www.ncbi.nlm.nih.gov/pmc/articles/PMC5400179/.

6
Infection

6.1 Introduction

In order to assure the biocompatibility of implants, biomedical biomaterial scientists and engineers must address not only aspects which promote normal wound healing and new functional tissue formation in conjunction with implants, but also must address how to avoid and/or minimize consequences which are undesirable. Detrimental outcomes jeopardize the fate of implants, and, most importantly, they cause illness as well as endanger the welfare (and sometimes the lives) of recipient patients. A pertinent, critical clinical complication is infection.

Infection is a pathological condition which results from the invasion and multiplication of pathogenic microorganisms (such as bacteria, viruses, fungi, etc.) in tissues, organs, and around implants inside the human body. The aforementioned pathogens are parasites which, in order to survive, must enter and live inside organisms of other species (e.g., humans), benefiting from the host milieu of satisfactory temperature, pH, and nutrient conditions.

The toxins and other metabolic chemicals produced by invader pathogens during their lifespan can cause fever and other health-related problems to hosts. The immune system in healthy human bodies fights infection-induced illnesses. Overcoming pathological complications is assisted with the help of appropriate medications prescribed by physicians. When pathogens are not controlled or annihilated, infection sets in; the health-related consequences can be serious and detrimental for inflicted patients.

In order to survive, pathogenic microorganisms have developed several ingenious, resourceful, and effective ways for protected survival and growth inside the host bodies. One of those means is the biofilm.

Concepts of Tissue-Biomaterial Interactions: Fundamentals and New Directions,
First Edition. Rena Bizios and Marissa E. Wechsler.

Companion Website: www.wiley.com/go/Concepts_of_Tissue-Biomaterial_Interaction_1e

6.1.1 Introduction to Biofilms

By definition, a *biofilm* is a multicellular community or colony formed by pathogenic microorganisms (such as bacteria, fungi, etc.) enclosed inside a protective cover (colloquially referred to as "slime", because it looks exactly like that) [1].

An example of a biofilm is the dental plaque, a slimy buildup produced by bacteria in the mouth, which is formed on the surface of teeth and is treatable in dentistry. Other biofilms, and their potential complications, have been of interest to health-care personnel because they are associated with infections. Examples of biofilm-related diseases and clinical complications include chronic lung diseases (such as cystic fibrosis), ear infections, various wounds, and, in addition, complications related to implants such as prosthetic bone implants, cardiac valves, catheters, etc. In all aforementioned cases, more than 80% of the microbial infections are associated with biofilm formation [2, 3]; it should be noted that these types of infections are difficult to diagnose and treat in infected patients.

6.1.2 The Process of Biofilm Formation

The cell communities contained inside the biofilm are composed of either one and the same or different types of pathogenic microorganisms dispersed throughout the extracellular matrix they produce. Such extracellular matrix contains proteins, polysaccharides, and several other chemical compounds. Examples of bacteria which form biofilms include *Escherichia coli* (some of which are benign and are found in the human gastrointestinal system, but some strains are pathogenic), *Staphylococcus aureus* (a microorganism which survives, and is propagated, in air), *Staphylococcus epidermidis* (present on skin), and *Salmonella enterica* (which is found in the intestines) [4].

Formation and the life cycle of a biofilm involve several steps (Figure 6.1). At the start, single planktonic bacteria swim in body liquids propelled with the assistance of their characteristic tail, known as "flagellum" (Figure 6.1A), slow down when they encounter a substrate, and first form transient and reversible interactions with substrate surfaces (Figure 6.1B). Within seconds to minutes, however, planktonic bacteria irreversibly adhere to substrate surfaces and lose their flagella (Figure 6.1B). Within hours to days, the adherent bacteria proliferate and form a colony, which grows as a result of cell proliferation and synthesis and deposition of various chemical compounds in their extracellular matrix (Figure 6.1C). The result is formation of a three-dimensional biofilm (Figure 6.1D). As time progresses from days to months, the bacteria colony becomes bigger because it

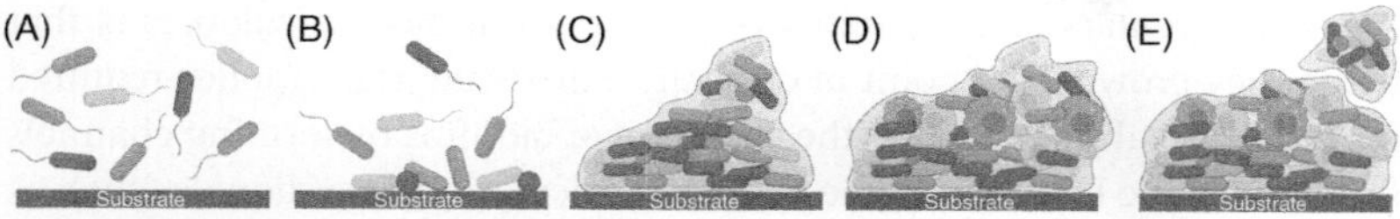

Figure 6.1 Schematic illustration (not to scale) of biofilm formation. (A) Planktonic bacteria (each with a characteristic flagellum) flow above a substrate. (B) Some motile bacteria interact with the substrate and then permanently adhere, each losing its flagellum. (C) Bacteria proliferation results in colony formation. Subsequent development of a biofilm encloses and protects the colony. (D) The bacteria colony inside the biofilm grows as a result of proliferation and entrapment of other cells and debris. (E) When the original colony grows too large to exist, parts may break off, flow downstream, adhere to another host site, and form a new pathogenic colony.

incorporates not only more bacteria but also various other cell types which, as they are flowing by, are trapped and are incorporated within the biofilm (Figure 6.1D).

Bacteria, which are the constituent cells inside biofilms, attach not only to substrates but also to each other with the help of the extracellular matrix components, which these bacteria produce and deposit on the surfaces they colonize [4]. Inside biofilms, bacteria use chemical compounds (such as specific proteins and carbohydrates, which they are programmed to synthesize and release) to efficiently communicate with each other. The genetic expression by bacteria inside biofilms is different from that of the planktonic bacteria (that is, cells prior to their irreversible attachment to substrate surfaces).

Biofilms are complex environments which evolve and change with respect to time. Once the biofilm colony is formed, the dwelling bacteria specialize to perform various functions needed for their survival and growth of the biofilm [4]. In this respect, some bacteria are capable of synthesizing the biofilm. Other bacteria contained in biofilms may become one of the following: "fixed cells", which permanently reside in the biofilm; "motile cells", which have the capability to leave the colony, go further downstream, and start a new colony at other parts of the host body (Figure 6.1E); "pillar cells", which pile up and form three-dimensional arrangements resembling pillars; and cells in permanent monolayer arrangement. The overall arrangements are multilayered, three-dimensional, microorganism-containing structures [4].

An important aspect, which must be emphasized, is that the enclosed biofilm environment provides protection and also maintains the conditions (such as nutrients) required for the survival and growth of bacteria communities, even under dynamic environments (e.g., bulk flow conditions).

Another remarkable capability of pathogens in biofilm colonies is that when they grow to the point of critically exhausting the supplies required for cell survival, they modify the structure of biofilms by forming channels throughout the biofilm volume and, thus, increase flow of liquids (such as blood), which contain growth nutrients and oxygen – the critically needed components for survival and growth of bacteria colonies in biofilms. In addition, starvation conditions at originally infected sites induce formation of planktonic (i.e., motile) cells. Following detachment, these cells migrate to other parts of the infected organism, attach to other substrates (natural and synthetic), grow new bacteria colonies, and form more biofilms (Figure 6.1E). These events result in systemic and chronic infection conditions, which are of critical importance for the welfare of patients.

6.2 Clinical Treatments of Infection

The fight against infection in clinics and hospitals starts by carefully and thoroughly cleaning wound sites; in the case of skin wounds, chemical compounds (such as hydrogen peroxide and alcohol) are used to remove extraneous debris (for example, dirt, *etc.*) and minimize possibilities of infection. Subsequent surgical interventions must take place under sterile conditions using sterile materials and sterile, surgical instruments.

Different types of wounds may show different signs of infection. All infected wounds, however, exhibit some of the following signs: fever, pus, abnormal smell, warmth and redness, delayed healing, and edema. Among these symptoms, warmth/fever, redness, and edema are the cardinal signs of inflammation (reviewed in Chapter 2: Section 2.5).

Appropriately trained and specializing biomedical personnel use various microbiology techniques to identify the type of pathogen(s) involved in infection-causing diseases. *In vitro* pathogen cell cultures are also used to determine which among existing medications (such as antibiotics) annihilates the infection-causing pathogen(s). Based on the pathogen identification results, attending physicians choose the medication(s) needed to treat and, thus, help suffering patients.

Furthermore, the aforementioned *in vitro* pathogenic cell cultures are used not only to identify pathogens resistant to currently available medications but also to establish the efficacy of novel medications developed to combat the resistant pathogens. Diagnoses of pathogen strains resistant to currently available medications are of paramount importance in the case of chronic infections. Unique challenges are the cases of infected wounds which are chronic and do not heal. These wounds are associated with certain pathological conditions, such as ulcers on the feet and toes of diabetic patients, sores on the back of chronically immobile patients suffering from paralysis and resting in one body position for prolonged periods of time, and other serious incapacitating pathological conditions.

6.2.1 Medical Treatments of Infection

Health-care personnel fight infections by using and delivering medications (such as antibiotics to treat bacteria-related health problems). Antibiotics in the form of pills are taken by patients primarily through gastrointestinal delivery. Antibiotics, which are available only in liquid form, are injected either intramuscularly or intravenously. Biomedical engineers and scientists have designed and developed other specific approaches pertinent to drug delivery; for example, they incorporated antibiotics into implantable, synthetic biomaterials for subsequent release either from the surfaces of non-degradable materials or from materials as they degrade post implantation.

Although antibiotics proved to be a most effective way to fight bacteria-caused infections, they are not the panacea, and they do not solve all the infection-related problems. Complications arise because various infections are caused by different types of bacteria, which may also have different sensitivities to antibiotics that are available and are prescribed by doctors to treat infected patients. Other limitations regarding the mode of antibiotic delivery to specific infected sites, for example, to tissues around infected implants, include the duration of medication release, extent of diffusion (i.e., distance from the source of antibiotics to the infected site), and release of effective antibiotic dose post implantation. Another aspect which may affect the efficacy of antibiotics is the biochemical milieu of the host body, which contains many bioactive chemical compounds and involves many chemical reactions. Such bioactive chemicals may minimize, and even neutralize, the efficacy of prescribed medication(s) post delivery.

Last but not least, antibiotics may trigger unwanted sensitivities and allergic reactions; for example, some patients are critically sensitive to penicillin. When penicillin was first discovered, it was received with great excitement because it treated effectively several untreatable diseases of those times (e.g., tuberculosis). Penicillin alone is not prescribed anymore because this antibiotic is not effective in annihilating the strains of pathogens infecting patients nowadays. Furthermore, some patients develop allergic reactions (which range from mild to very serious life-threatening responses) to certain medications. When known, such sensitivities are taken into consideration by health-care personnel in choosing appropriate medication(s) for treating each patient.

Since the discovery of penicillin, various other antibiotics produced and marketed by pharmaceutical companies have been used to treat patients. As time progressed, however, many antibiotics became ineffective in treating pathological conditions because of the emergence of virulent strains of pathogens, which proved resistant to existing antibiotics [4]. From the perspective of pathogens, mutations producing new variants/strains resistant to existing medications (e.g., antibiotics) is a most effective strategy pathogens have used in order to survive! Frequent and recurring emergence of antibiotic-resistant strains of pathogens is a major clinical problem because it renders existing medications ineffective in treating patients suffering from infected wounds and other infectious conditions. For this reason, novel medications are constantly needed. Availability of such novel medications, however, has almost stopped because pharmaceutical companies (the source of medications) choose instead to devote their research, formulation, and production toward new, more profitable medications for other syndromes.

Despite several important advances in the treatment of infections in the clinical milieu, certain challenges in fighting these pathological conditions still remain unresolved and need to be addressed. Specifically, the clinical need for new and effective antibiotics and other medications to treat infections and infectious diseases requires continuous search for novel, effective medications.

6.2.2 Infections Associated with Implants

The presence and growth of pathogenic microorganisms (such as bacteria and viruses, especially under the protective, enclosed environment of biofilms) have been of great interest to many people. Biomedical engineers are interested because the presence and growth of pathogenic microorganism colonies and biofilms, in conjunction with implant materials and medical devices *in vivo*, may not only result in patient-related health problems but also cause implant failure with serious consequences: either leading to implant rejection or requiring surgical excision.

Biomedical scientists and engineers discovered and advocated that, when infection is established at the specific anatomical location where biomaterials and medical devices are implanted, the clinical outcome depends on the competition between the invading pathogens (determined to survive by attaching, proliferating, and functioning on and around implants) and the ability of the recipient host to defend itself through the protective mechanisms of the immune system, the natural defense system of the human body. Major participants in this process are activated white blood cells, which are programmed by nature to recognize the invader pathogens, engulf them in their cytoplasm, and annihilate them via phagocytosis. The competition between pathogens and mammalian cells (from the surrounding normal tissues) as they strive to arrive, interact with, and function for dominance of the same implant surface is referred to as "The Race for the Surface" [5, 6]. Undoubtedly, the outcome of that race for the implant surface is crucial for the welfare of patients and for the fate of implants *in vivo*.

6.2.3 Implant Material Properties Contributing to Infection

Infections post implantation associated with biomaterials and implant devices, such as hip prostheses, prosthetic heart valves, intrauterine devices, urinary catheters, and central venous catheters, have been reported in the literature. Excision has become an unavoidable medical care decision because infected implants cause pain, malfunction, and other clinical complications jeopardizing the welfare of recipient patients. Following excision of infected implants, microbiological analysis revealed that (i) infection inflicts implants in both soft and hard tissues in all parts of the human body and (ii) several types of pathogens infect and form biofilms on implants. Such pathogens include *Candida albicans*, *Coagulase-negative staphylococci*, *Enterococcus* spp., *Pseudomonas aeruginosa*, and *Staphylococcus aureus* [1].

Biomedical engineering research identified several parameters responsible for pathogen interactions with, and formation of biofilms on, biomaterial surfaces. From the material perspective, substrate properties, specifically texture, roughness, and hydrophobicity, as well as the presence and type of adsorbed chemical compounds (such as adhesive proteins), are contributing parameters. An important aspect which should never be forgotten is the fact that, within seconds upon their introduction in the body, implanted biomaterials and medical devices are covered with absorbed proteins which are present in body liquids and in surrounding tissues [7]. From the perspective of pathogens, pertinent aspects include the pathogen cell membrane hydrophobicity, absence of the flagella on bacteria (when they become adhesive), and formation of the extracellular polymeric substance (i.e., the extracellular matrix produced and deposited by the pathogens). When implants are in an environment which involves liquid flow, the parameters affecting attachment of the pathogenic microorganisms on implant material surfaces include liquid flow velocity, pH, temperature, as well as the presence of cations and antimicrobial agents [1].

Invaluable facts garnered from extensive research regarding infections associated with implants are that (i) bacteria growing in the presence of implants are tenacious and difficult to deal with prescribed medications (for example, antibiotics), (ii) chronic infections require surgical removal of implants, and (iii) reimplantation at the same anatomical location (*if it is possible*) must be delayed until the infection is successfully eliminated and the surrounding tissues are healed.

6.2.4 Potential Strategies to Overcome Implant-Associated Infections

Lessons learned from research regarding interactions of pathogens with material substrates provide valuable information regarding potential targets for attacking the process of biofilm formation on implants in order to eliminate relevant clinical complications post implantation.

For this purpose, knowledge and understanding of the vulnerability during pathogen interactions with, and formation of biofilms on, substrates are instructive and can become valuable strategies. One potential target is the early planktonic cell adhesion on material surfaces; medications (such as antibiotics) are effective against planktonic (versus later strains) of pathogens. Other strategies could focus on the formation of biofilms. From the biomaterial perspective, therefore, some strategies which are very promising in defeating pathogen-related infections and growth of biofilms are (i) to design, formulate, and use biomaterials which resist adhesion of pathogens, (ii) design, formulate, and use novel biomaterials which incorporate and subsequently release pertinent medications known for effectively targeting pathogens, and (iii) to disrupt initiation and formation of mature biofilms [4].

6.2.5 Current Research Endeavors to Overcome Bacterial Infections of Implanted Devices

Other approaches, targeting pathogen-related infections associated with implanted medical devices, focused on the diagnosis of infection. In this respect, biosensors have been developed and used for early detection of infections. Advanced imaging techniques and nanoparticles were used for detecting and localizing sites of infections *in vivo*.

Antimicrobial photodynamic therapies used oxygen radicals (produced from reaction of light and oxygen in photosynthesizers) to disrupt pathogen colonies. Bioacoustic techniques (such as ultrasonication) were used to shake up biofilms in order to either disrupt or break up pathogen cell adhesion on material surfaces. Electric and magnetic fields were used to detrimentally affect survival and function of pathogenic microorganisms.

Last but not least, techniques to break up and dissolve the extracellular polymeric substances (which had been deposited by pathogens in their colonies) were investigated.

Approaches using chemical compounds to prevent pathogen interactions with material surfaces were also explored. For this purpose, antimicrobial chemical compounds (such as antibiotics) were coated onto material surfaces in order to kill bacteria upon first contact with such implants *in vivo*. Other approaches focused on disrupting biofilm formation using various chemical modifications on biomaterial surfaces in order to prevent subsequent interactions with pathogenic microorganisms [8].

6.3 Future Directions: Challenges and Opportunities

Future directions addressing challenges as well as pursuing pertinent opportunities regarding biomaterials are to first design, formulate, and produce infection-resistant materials and then to design, formulate, and produce implantable biomaterials and medical devices which both promote surrounding mammalian cell–tissue interactions (pertinent to formation of new, functional tissue) and, at the same time, resist infections by preventing interactions with pathogenic microorganisms.

Undoubtedly, there is a great need for novel, creative ideas which future generations of biomedical engineers should explore and successfully implement. Such novel approaches and products are needed to help patients who are suffering from various pathological conditions resulting from infections around implanted biomaterials and medical devices.

References

1 Donlan RM. Biofilms: microbial life on surfaces. *Emerg. Infect. Dis.* [Internet]. 2002 [cited 2023 Aug 29];8(9):881–90. Available from: https://pubmed.ncbi.nlm.nih.gov/12194761/.

2 Jamal M, Ahmad W, Andleeb S, Jalil F, Imran M, Nawaz MA, et al. Bacterial biofilm and associated infections. *J. Chin. Med. Assoc.* [Internet]. 2018 [cited 2023 Aug 29];81(1):7–11. Available from: https://pubmed.ncbi.nlm.nih.gov/29042186/.

3 Vishwakarma A, Dang F, Ferrell A, Barton HA, Joy A. Peptidomimetic polyurethanes inhibit bacterial biofilm formation and disrupt surface established biofilms. *J. Am. Chem. Soc.* [Internet]. 2021 [cited 2023

Aug 29];143(25):9440–9. Available from: https://pubs.acs.org/doi/abs/10.1021/jacs.1c02324.

4 Arciola CR, Campoccia D, Montanaro L. Implant infections: adhesion, biofilm formation and immune evasion. *Nat. Rev. Microbiol.* 2018 [cited 2023 Aug 29];16(7):397–409. Available from: https://www.nature.com/articles/s41579-018-0019-y.

5 Ratner BD, Hoffman AS, Schoen FJ, Lemons JE. *Biomaterials Science An Introduction to Materials in Medicine.* 2nd ed. Elsevier; 2004.

6 Black J. *Biological Performance of Materials: Fundamentals of Biocompatibility.* 2nd ed. New York, NY: Marcel Dekker Inc.; 1992.

7 Ratner BD. The catastrophe revisited: blood compatibility in the 21st Century. *Biomaterials* [Internet]. 2007 [cited 2023 Aug 29];28(34):5144–7. Available from: https://pubmed.ncbi.nlm.nih.gov/17689608/.

8 Khatoon Z, McTiernan CD, Suuronen EJ, Mah TF, Alarcon EI. Bacterial biofilm formation on implantable devices and approaches to its treatment and prevention. *Heliyon* 2018;4(12):e01067.

7

Sterile Conditions. Sterilization of Implants

7.1 Introduction

Over the centuries of human history, pathogenic microorganisms have caused major medical problems, including epidemics and pandemics. For a long time, it was not known what caused such major medical problems because even the existence of microorganisms was not known. Advances in instrumentation (specifically powerful microscopes) provided visualization of microorganisms, followed by research regarding pathogens, which resulted in the treatment of inflicted patients. An example of a pertinent development in fighting pathogens is, of course, the discovery of antibiotics.

Case Study

What did people do in the past and before the "age of antibiotics"? In this respect, there are anecdotal cases of successful elimination of pathogens from sources such as contaminated water supplies, which have the potential for inflicting entire communities and causing major medical problems, including patient death. One such case was published in *Science* [1].

This case is an example of old, "traditional" disinfection, and it was entitled "Life Saving Pitchers." Somewhere in rural India, people believed that using brass pitchers to store water (obtained from a nearby river) for drinking purposes prevented illness. Scientists from a University in the United Kingdom in collaboration with colleagues from a University in India decided to do an experiment and scientifically determine what was the reason for such a desirable outcome.

(Continued)

Concepts of Tissue-Biomaterial Interactions: Fundamentals and New Directions, First Edition. Rena Bizios and Marissa E. Wechsler.

Companion Website: www.wiley.com/go/Concepts_of_Tissue-Biomaterial_Interaction_1e

(Continued)

The experiment conducted used brass pitchers (which people in rural India used) as well as other earthenware and plastic water containers. In one group, the pitchers were filled with sterile water, but, on purpose, the researchers added *Escherichia coli* (a pathogenic microorganism) to them. In the other group, all pitchers (i.e., brass, earthenware, and plastic) contained water from the river (the source of contaminated drinking water used by the Indian villagers) [1].

The results of this experiment were as follows: the bacteria count in the brass pitchers dropped from starting as high as 1,000,000 per milliliter to zero after two days; in contrast, the bacteria levels in the earthenware and plastic pitchers remained very high during the same time period. The explanation provided by the researchers for this result was that the brass pitchers released copper ions, which are known for killing various cells, including bacteria. Thus, this study provided a scientific explanation regarding the mechanism underlying the protection which the brass pitchers provided to villagers in rural India for centuries: the small number of released copper ions killed the bacteria when the contaminated river water was stored in brass pitchers overnight without causing other medical problems. The small number of released copper ions was a fortuitous coincidence because in high concentrations, copper ions are very toxic; for this reason, copper is not used for metal implants.

7.2 Determining Implant-Related Sterile Conditions

Prevention of the contamination and infection of biomaterials and medical devices implanted into the human body and the bodies of animals is a paramount medical requirement because if infection sets in around implants, it may cause serious illnesses and may even lead to the death of patients who had received implants that became infected. In order to avoid such undesirable outcomes and medical complications, implants must be sterile before, and remain sterile post implantation in patients. This requirement is an integral part of all aspects of biomedical implant production (i.e., manufacturing, packaging, transport, delivery to hospitals, etc.) and has become the responsibility of production engineers in companies specializing in implants. The aforementioned requirements are met by establishing, using, and maintaining sterile conditions at all stages of implant production and subsequent use in patients.

By definition, *sterile* is a condition characterized by the absence of *all* living organisms (such as bacteria, yeast, mold, and viruses). An important distinction must be made: "sterility" is different from "cleanliness".

In conjunction with implants, sterile conditions are attained following established procedures. Experiments and pertinent tests are performed in order to establish the sterility of implant biomaterials and medical devices. This process starts with determination of the presence, number, and type of the contaminating microorganism(s). For that purpose, specimens of the material under consideration are each immersed in clear, sterile liquid (e.g., cell-culture medium, which promotes cell survival and proliferation) at 37 °C (simulating the average human body temperature), in a humid environment for several hours (e.g., 24 hours). At the end of that time period, the turbidity of the supernatant liquid from each material specimen is checked. In case of microbial growth, the starting clear liquid becomes turbid; this is evidence that the original material specimen was contaminated by microorganisms. The pathogens causing the turbidity observed in the aforementioned test are collected and further examined to identify the microorganism type(s) involved. The aforementioned tests require microbiology techniques established and used in specializing microbiology laboratories. The results provide information regarding the *bioburden*, that is, the number of viable microorganisms on the implant biomaterial and medical device prior to sterilization [2].

Another important aspect that must be determined when sterilizing implants is the duration of exposure to the sterilization process necessary to achieve the desirable outcome, specifically elimination of pathogens on the tested implants. Once the sterilization process is completed, the criterion to determine that sterility had been achieved is the *Sterility Assurance Level* (abbreviated SAL). Sterility Assurance Level is the probability that a given implant will remain nonsterile following exposure to a sterilization process. The accepted maximum SAL value for implants is 10^{-6}.

The implant device bioburden is determined prior to choosing and adopting a sterilization method. First, the number of adhering microorganisms on the material specimens tested is determined (and designated as the initial number of the adhering microorganism). The bioburden on several material specimens (for example, 10–30 such samples) is then determined by performing fractional-run sterilization experiments. For this purpose, each material sample is exposed to the chosen sterilization method for various periods of time. At the end of each prescribed time point, the adhering microorganisms on each tested material sample are rinsed and shaken off; each effluent (containing the adherent microorganisms) is collected into a small volume of a sterile recovery liquid (for example, an appropriate cell-culture medium). The number of viable microorganisms remaining after material sterilization for each specific period of time is determined. As the microorganism-carrying implant is exposed to the process of sterilization for prolonged periods of time, the microorganisms are killed, and the number of viable pathogens decreases as a function of time.

The collected data are plotted as follows: the log of the number of remaining viable microorganisms (on the y-axis) versus the duration of the exposure of microorganisms to the sterilization process under consideration (on the x-axis; in minutes). A straight line with a negative slope is thus obtained. The intersection point of the diagonal, negative slope line and the horizontal line from the 10^{-6} (which is the SAL) point on the x-axis is the length of exposure time (expressed in minutes) needed in order to achieve implant sterilization using the process under consideration. This specific length of time is needed to achieve successful sterilization of the tested biomaterial [2].

7.3 Methods for Implant Sterilization

Another important consideration for successful sterilization of implants is selection of the appropriate sterilization method. First and foremost, the integrity of the implant biomaterials and medical devices must be maintained during, and following, the sterilization method used. The integrity of packages or any other enclosure (in which implants are contained during transport from the factory to the clinics and hospitals) must be maintained at all times. After sterilization, sterility must be maintained at all stages of handling, transport, and storing until the sterile implants are used in surgical procedures at hospitals and clinics.

The methods currently used for sterilization of implantable biomaterials and medical devices are (i) autoclaving or steam sterilization, (ii) ethylene oxide (EtO) sterilization, and (iii) sterilization using radiation emitted from various isotope sources. Brief reviews for each one of these three methods of implant sterilization follow.

7.3.1 Autoclaving or Steam Sterilization

Autoclaving or steam sterilization is a widely used procedure that exposes implant materials and devices to saturated steam at 121 °C under pressure for approximately 15–30 minutes or until all surfaces of the implants reach a temperature of at least 121 °C [2].

The process of autoclaving or steam sterilization requires special equipment, which must withstand high-temperature and high-pressure conditions for the duration of the sterilization process. Such laboratory apparatuses are available in various sizes, from bench-top (0.5 m^3 in volume) models to walk-in rooms in hospitals, where a large number of surgical instruments, surgical gowns, etc., must be sterilized daily.

The effectiveness of autoclaving results from the detrimental impact that the combined high temperature and pressure have on cells; such exposure kills microorganisms by destroying their metabolic and structural components, which are needed for cell survival and proliferation. One of the affected underlying mechanisms involves coagulation of essential enzymes and another one disruption of protein and lipid complexes [2].

Autoclaving is currently used for sterilizing glass, metals, and liquids in research laboratories. This procedure is also used in hospitals for sterilizing metallic surgical instruments as well as heat-resistance surgical supplies such as linen, drapes, medical dressings, etc. Autoclaving is a popular and widely used sterilization method because of its advantages: it is relatively simple to use, fast, easy to control, and it is efficacious, giving the desirable result of eliminating infecting microorganisms without leaving any toxic residue. Due to the high-temperature and pressure conditions of the steam sterilization method, however, most nonmetallic implant devices and packaging materials (specifically polymers) cannot be sterilized using this method (because most polymers melt under high-temperature conditions). For this reason, other sterilization methods are used to sterilize polymeric biomaterials and medical devices.

7.3.2 Ethylene Oxide (EtO/EO) Sterilization

During the process of EtO sterilization, implant biomaterials and devices are exposed to EtO gas concentrations of 600–1200 mg/l, in relative humidity of 40–90%, at a temperature of 30–50 °C, for 2–28 hours [3].

This process must take place inside a specially manufactured sterilization chamber, separately housed facility, and operated under special conditions. The air originally contained in that chamber is removed before the EtO is injected. Once established, the EtO environment must be maintained for the duration of the sterilization procedure *without* leaks and disruptions. At the end of the procedure, any EtO gas remaining inside the sterilization chamber must be removed *very carefully* and without any leaks or accidents.

Ethylene oxide is a very effective sterilization method because it affects pathogens at their molecular level (specifically at the nucleic acid RNA and DNA levels) and is, therefore, lethal to microorganisms exposed to that gas. Use of the EtO sterilization method has several advantages. First and foremost, it has high penetration capability, which assures that all facets of three-dimensional structures of various implants are sterilized. Ethylene oxide sterilization is an efficacious method even at low temperature (its operating temperature is in the 30–50 °C range) [3].

Ethylene oxide sterilization is used successfully for various biomedical implants including surgical sutures, intraocular lenses, ligament- and tendon-repair devices, *in vivo* absorbable and nonabsorbable meshes, neurosurgery devices, *in vivo* absorbable bone repair devices, heart valves, vascular grafts, etc. It should be noted that many of the aforementioned medical devices are composed either entirely of polymers or of combinations of polymers with other material types. The ethylene oxide treatment is compatible because it sterilizes many types of materials (including polymers which are sensitive to high temperatures) without detrimental effects on material chemistry and properties (such as loss of structure integrity due to melting).

Use of the EtO method of sterilization, however, requires special precautions (including custom-built facilities and special operating arrangements) because EtO is a toxic chemical and a suspected carcinogen. Detrimental health-related effects can be caused by even few residue molecules of EtO gas clinging on implantable biomaterials and medical devices. Such residue remains clinging on material surfaces even after the sterilization process is completed. For this reason, extended shelf-time is needed to remove effectively the clinging EtO residue following exposure of implants to EtO sterilization. During that time period, the adhering EtO molecules are released from the sterilized implants inside hermetically sealed chambers designed specifically for that purpose. The extra shelf-time precaution is mandatory before the sterilized materials and devices are delivered to hospitals and are implanted in patients. In addition, EtO can have detrimental health-related effects on the personnel who work in the facility where EtO is used for sterilizing implants.

At the end of each EtO sterilization process, the remaining gas in the special sterilization chamber and facility cannot be released into either the atmosphere or the surroundings because of its potential toxicity and carcinogenicity. Last but not least, both pure EtO and mixtures of this gas (without a proven inert compound) are flammable and potentially explosive [4].

7.3.3 Sterilization Using Emitted Radiation

Another method for sterilizing biomaterials and devices is using radiation emitted from a radioactive isotope, such as Cobalt-60 (60Co). The radiation sterilization procedure involves exposure of implant materials and medical devices to gamma rays released from an isotope source such as 60Co at a dose of 25 kGy (*Note*: Gy = Gray = 1 Jule/kg) [3, 5, 6].

The pertinent operation requires a special irradiator machine and a facility where radioisotopes are stored and used. The biological effects of radiation sterilization are well documented: gamma rays have high penetrating ability *but only in the linear mode.* Gamma rays ionize key cellular components (such as nucleic acids) and, thus, kill microorganisms.

The advantages of using radiation sterilization are that it is a simple, fast, and effective method which does not leave any toxic residue on the surfaces of the sterilized biomaterials and medical devices. This method is compatible with some but not with all polymers (specifically polyethylene, polyesters, polystyrene, polysulfones, and polycarbonates). Biomedical biomaterial scientists and engineers should be aware of this compatibility limitation before exposure of vulnerable polymers to radiation sterilization. The detrimental effects of exposure to radiation involve breakdown of the polymer backbone, causing unexpected changes in the material chemical composition and topography and subsequent *in vivo* malfunction of implants, leading to post-implantation complications in patients.

Other disadvantages of the radiation sterilization process include the cost required for building a special facility, installing an isotope, and running that facility for implant sterilization purposes. Due to their continuous decay, the isotopes used for this sterilization process must be replaced at regular time intervals. In addition to the needed funds and appropriate expenditures to obtain and install a new isotope, the entire facility must be shut down for a period of time for maintenance, replacement of isotope, etc. In order to avoid delayed delivery of implants to customers in hospitals and clinics, the biomedical engineers (who work in industry and are responsible for sterile product delivery) must be cognizant of such delays and must make appropriate schedule plans to assure smooth production and product delivery to the medical customers.

Last but not least, extra care is required to ensure that the radiation sterilization facilities are a safe environment for the personnel who work there. Exposure of personnel to radiation must be avoided because of potential health-related consequences and problems.

7.3.4 Development of New Sterilization Methods

In addition to the aforementioned three methods of sterilization for implant biomaterials and medical devices, several other emerging technologies are being considered, experimented with, and developed. These endeavors include gaseous chlorine dioxide, low-temperature gas plasma (i.e., a gaseous environment containing ionized gases), gaseous ozone (which contains oxygen radicals known for their biochemical activity in killing microorganisms), machine-generated X-rays, and vapor-phase hydrogen peroxide (a gas which, at room temperature, is very effective in decontaminating entire rooms (such as laboratories and all contained furniture) without any detrimental effects on personnel, instruments, and materials) [3, 7].

The continued (and in many occasions) important need for sterilized materials and devices (which either come in contact with or are implanted in the human body), as well as for sterile conditions in surgery rooms, hospitals, clinics, and cell-culture laboratories, justifies research in relevant fields and has the potential for new developments and contributions in the future.

7.4 Biomaterials: The Biomedical Engineering Connection

Sterilization of implant biomaterials and medical devices is a responsibility of biomedical engineers working in pertinent industries. Companies (which are providers of implantable materials and medical devices) must deliver sterile products to hospitals and clinics. For this reason, such products are sterilized in the companies prior to packaging and must remain sterile during handling, transport, and delivery to the destination hospitals and clinics.

How are the sterile conditions maintained afterwards? These aspects are the responsibility of hospitals, clinics, surgery rooms, intensive-care units (where these products are subsequently used in conjunction with patients), and the patient-attending healthcare personnel.

7.4.1 Sterile Conditions and Practices in Surgery Rooms and Hospitals

Following delivery to hospitals and clinics, the supplied sterile biomaterials and medical devices must be first stored, and subsequently used, under conditions which do not compromise sterility. For this purpose, certain precautions must be, and are taken, at the sites where pertinent implant procedures are performed.

The surgery rooms in hospitals are built using special, nonporous building materials, which (at regular time intervals) can be sterilized using appropriate solutions of pathogen-killing chemicals (such as bleach and alcohol). Surgical beds and all other furniture, fixtures and equipment in surgical rooms are made of materials (such as metals, textiles, etc.), which can be successfully sterilized regularly. The temperature in surgery rooms is kept low; this condition prevents growth of microorganisms (which proliferate much faster in warm environments). The surgery rooms have special ventilation systems directing the circulating air away from the surgical bed (on which the patient lies during surgery) in order to eliminate the possibility of infecting open wounds with airborne pathogens from the air circulating inside hospitals.

Before surgery, all medical personnel participating in surgical procedures wear sterile gowns, gloves, hair covers, face masks, etc., and "scrub"; that is, they wash their hands and arms thoroughly using appropriate soap (for example, chlorhexidine gluconate) and special disinfectants (which are chemicals known to kill microorganisms); then, they air-dry their clean hands and arms before wearing sterile gloves. During surgery, the surgeons and all assistants wear sterile gowns, masks, gloves, hair covers, shoe covers, etc. Only sterile instruments are used during surgery. Moreover, all surgery instruments are used once and for only one patient [4, 8].

Before surgery, the skin of the patient (at the anatomical site of the intended surgery) is "prepped", that is, topically sterilized using povidone-iodine (e.g., betadine, a commercially available product), which is a disinfecting chemical solution known to kill microorganisms. Before, as well as after, surgery, patients are given pertinent antibiotics and other prescribed medications to prevent infection and related health complications [4, 8].

Taken together, the aforementioned procedures used in the surgery rooms of hospitals and clinics establish, and maintain, a sterile milieu in order to prevent (and hopefully eliminate) the potential of infecting patients [4, 8].

References

1 Kaiser J. Lifesaving pitchers. *Science* [Internet]. 2005 [cited 2023 Aug 31];308(5719):195–195. Available from: https://www.science.org/doi/10.1126/science.308.5719.195a.

2 Harrington RE, Guda T, Lambert B, Martin J, 3.1.4 – Sterilization and disinfection of biomaterials for medical devices. WR Wagner, SE Sakiyama-Elbert, G Zhang, MJ Yaszemski, editors. *Biomaterials Science.*

4th ed. Academic Press; 2020, pages 1431–1446. ISBN 9780128161371. https://doi.org/10.1016/B978-0-12-816137-1.00091-X.

3 Dai Z, Ronholm J, Tian Y, Sethi B, Cao X. Sterilization techniques for biodegradable scaffolds in tissue engineering applications. *J. Tissue Eng.* [Internet]. 2016 [cited 2023 Aug 31];7. Available from: https://pubmed.ncbi.nlm.nih.gov/27247758/.

4 Rutala WA, Weber DJ. Disinfection and sterilization in health care facilities: an overview and current issues. *Infect. Dis. Clin. N. Am.* [Internet]. 2016 [cited 2023 Aug 31];30(3):609. Available from: https://www.ncbi.nlm.nih.gov/pmc/articles/PMC7134755/.

5 Haim Zada M, Kumar A, Elmalak O, Mechrez G, Domb AJ. Effect of ethylene oxide and gamma (γ-) sterilization on the properties of a PLCL polymer material in balloon implants. *ACS Omega* [Internet]. 2019 [cited 2023 Aug 31];4(25):21319–26. Available from: https://pubs.acs.org/doi/full/10.1021/acsomega.9b02889.

6 Nguyen H, Morgan DAF, Forwood MR. Sterilization of allograft bone: effects of gamma irradiation on allograft biology and biomechanics. *Cell Tissue Bank.* [Internet]. 2007 [cited 2023 Aug 31];8(2):93–105. Available from: https://link.springer.com/article/10.1007/s10561-006-9020-1.

7 Bento CSA, Gaspar MC, Coimbra P, de Sousa HC, Braga MEM. A review of conventional and emerging technologies for hydrogels sterilization. *Int. J. Pharm.* 2023;634:122671.

8 U.S. Department of Health and Human Services. Centers for Disease Control and Prevention (CDC). Guidelines for Environmental Infection Control in Health-Care Facilities Recommendations of CDC and the Healthcare Infection Control Practices Advisory Committee (HICPAC), 2019.

8

Evaluation of Biocompatibility

In the past, biomedical applications used materials that either existed or had been developed for other purposes, which were not specific to biomedical needs and did not meet exclusive implant specifications (as discussed in Chapter 1). Failure of those trial-and-error approaches necessitated and prompted continued endeavors for alternatives and improved biomaterials. For this purpose, biomaterial engineers and scientists have applied the lessons learned from Physiology (Chapter 2), Pathology (Chapter 3), and Biology (Chapter 4) in designing, formulating, and producing novel biomaterials for various applications.

Use of implant biomaterials and medical devices entails adherence to unique and special requirements, regulations, and laws: implants must always be safe for both the recipient patients and the health-care providers. In addition, implants must comply with regulations of various government regulatory agencies (such as the Food and Drug Administration (FDA) in the United States). For these reasons, successful design, formulation, and production of implantable biomaterials and medical devices require subsequent determination of their *biocompatibility* by providing evidence of successful performance under conditions pertinent to end-use application in the physiological milieu. Moreover, evaluation of the biocompatibility of implant materials and medical devices is mandatory and must meet important regulatory requirements. Failure to do so, has serious medical and legal consequences.

In order to achieve the pertinent objective, a multistage process for determining the biocompatibility of implantable materials and medical devices has been developed and used. Such determination involves a sequence of both *in vitro* and *in vivo* tests, which have been developed with contributions from physicians, surgeons, health-care practitioners, biomaterial scientists,

Concepts of Tissue-Biomaterial Interactions: Fundamentals and New Directions,
First Edition. Rena Bizios and Marissa E. Wechsler.

Companion Website: www.wiley.com/go/Concepts_of_Tissue-Biomaterial_Interaction_1e

and engineers and by taking into consideration the requirements of various regulatory agencies.

The ultimate desirable objective of all implants is to succeed in promoting and supporting long-term normal tissue regeneration and formation of functional tissue surrounding implants *in vivo* without any adverse effects. Successful completion of the biocompatibility evaluation process for implantable materials and medical devices is required in order to receive approval by government, regulatory agencies, reach the medical markets, and thus become available to help patients.

8.1 Mechanical Properties of Materials

Before any biocompatibility-related evaluation is attempted, the material and mechanical properties pertinent to the implant under consideration should be tested *in vitro*. In order to characterize their functional performance, prototypes of the various devices (composed of synthetic materials) are designed, assembled, and used in several *in vitro* tests, which must be successfully completed before any biocompatibility tests are planned and conducted.

8.1.1 Mechanical Tests *In Vitro*

During mechanical testing *in vitro*, specimens of the material under consideration are subjected to uniaxial loading at controlled amplitude and rate either once or repeatedly (as in the case of fatigue-related tests). Pertinent mechanical tests are tensile, compressive, bending, creep, stress relaxation, torsion, and material stiffness-related tests [1]. Collected data are plotted in stress versus strain plots and stress–strain versus time plots. Based on these analyses and plots, parameters pertinent to the mechanical characteristics of the tested material such as moduli, yield strength, tensile strength, resilience, toughness of the material, and failure are determined.

It should be noted that the material specimens tested for their mechanical properties are destroyed during some of the aforementioned processes. In addition, these tests require availability of special instruments and use of pertinent protocols. Undoubtedly, mechanical type tests are needed in order to determine the suitability of materials for biomedical applications, especially applications which expose implants (e.g., orthopedic and dental ones) to mechanical loading conditions *in vivo* [1].

8.1.2 Material Tests *In Vitro*. Determination of Material Degradation

The pertinent prerequisite tests also include *in vitro* material-related tests. A very important aspect, which must be carefully considered and tested, is the post-implantation degradation potential of biomaterials in the physiological environment of the human body. Uncontrolled material degradation post implantation is undesirable and should be avoided. In some cases, however, controlled degradation is incorporated into the design and performance of implantable materials for specific purposes (when, for example, biodegradable materials are used in tissue engineering and in controlled drug-release applications). When degradation is part of the performance of an implant, then this material-related aspect must also be appropriately tested in order to assure that the designated purpose of the material is satisfied [1]. Material degradation post implantation, however, is not desirable for implants and should be avoided.

Quantification of biomaterial degradation follows well-established protocols and procedures. For some *in vitro* studies, material samples of specific dimensions and known weight are immersed in simulated body liquid and maintained at 37 °C for various periods of time. At prescribed time points, aliquots of the supernatant liquid are collected and analyzed for the presence and composition of chemical compounds (such as ions from metals and monomers from polymers) which were released from the specimens under consideration. At the end of prescribed time points, each material sample is inspected visually and compared to respective controls (maintained in a buffer but similar experimental conditions) for signs of degradation, for example, color and shape change, appearance of cracks, etc. Moreover, at each prescribed time point at the end of the experiment, each material specimen is weighed to determine mass loss during the experiment. At the end of the experiment, select physical and chemical material properties are measured and compared to the values of the respective property values for each sample before *in vitro* testing.

In the case of implantable metals, the process of degradation is *corrosion* resulting from oxidation-reduction chemical reactions in the physiological milieu. This process is affected by careless handling of metal specimens, causing cracks, crevices, and pits and, thus, changing the surface topography in ways that promote and enhance subsequent corrosion [1].

Corrosion of metals can also result from removal of the *passivating layer* on the metal surface by either mechanical or chemical means. A *passivating layer* (composed of adsorbed proteins on substrate surfaces) is formed within seconds when a material is introduced into the body and is exposed to body liquids (e.g., blood plasma), which contain several dissolved proteins (discussed in Chapter 2). When adsorbed albumin (the most abundant protein in blood plasma) covers a metal surface, it provides "protection" because the chemically active oxygen radicals (produced, for example, during oxidation–reduction chemical reactions) cannot reach the protected metal surface and induce corrosion. Degradation of metals is also affected by the biological/physiological environment of body liquids and tissues; various proteins, lipids, cells, bacteria, and other chemical compounds (released from activated blood and tissue cells as well as from bacteria) can either attach or adsorb on the surfaces of metal implants and trigger the process of corrosion.

Although ceramics are used in several biomedical applications, they are not recommended for applications under dynamic loading conditions (such as orthopedic and dental implants) because they crumble under mechanical loading [1]. These materials, however, are desirable because they promote interactions with bone cells.

A pertinent example of such an application is when the surface of metal implants (such as a hip prosthesis) is covered by a ceramic (for example, hydroxyapatite). The resultant composite implant material has two major advantages: the mechanical strength of the constituent metal to withstand mechanical loads (occurring at the anatomical site of the hip site post implantation) and, at the same time, a ceramic coating which is an attractive substrate for adhesion of osteoblasts from the surrounding bone and subsequent cell functions relevant to new bone formation.

In the physiological environment, ceramics degrade primarily as a result of water infiltration at the pH environment of the body. Moreover, ceramics are susceptible to failure via stress-induced degradation (physical disintegration) in a dynamic, mechanical environment. The rate of degradation of ceramics is controlled by the chemical susceptibility of the specific material (for example, hydroxyapatite, alumina, and titania). Another parameter affecting the rate of degradation of ceramics is the degree of their crystallinity and, last, but not least, the amount of water in their surroundings [1].

Many implantable devices are composed of polymers. Degradation of polymers results from swelling following water infiltration into the polymer bulk in the temperature, pH, etc., conditions of the physiological milieu. Polymers degrade because their carbon backbone chain breaks down due to hydrolysis in the aqueous physiological environment; water infiltration causes swelling; the chemical activity of reactive, free oxygen radicals causes oxidation. Last, but not least, the enzymatic reactions of biochemical compounds present in the physiological milieu can either cause or contribute to the degradation of polymers post implantation [1].

Because of the wide use of polymers in biomedical applications, they have been studied extensively. One pertinent aspect is polymer degradation via two types of *hydrolysis*. *Bulk Hydrolysis* involves water infiltration into the bulk causing polymer breakdown. In the case of *Surface Hydrolysis*, as the polymer is exposed to a water environment, it degrades gradually, layer by layer from the surface toward the center of the specimen. Such breakdown produces many water-soluble degradation products, including monomers.

In closing, a reminder regarding a major prerequisite requirement for the evaluation of the biocompatibility – specifically, successful, *in vitro* characterization of materials and of the functional performance of device prototypes – is needed before initiation of the evaluation of the biocompatibility of any implant material and device. If materials and devices do not pass the prerequisite *in vitro* tests, then they should not be evaluated for their biocompatibility; in other words, any plans for biocompatibility evaluation are abandoned. Biomaterial scientists and engineers must then start designing, formulating, producing, and characterizing novel biomaterials for implant-related applications.

8.2 *In Vitro* Hemocompatibility (Blood Compatibility) Tests

A special category for biocompatibility evaluation is materials which will be used for cardiovascular applications. These tests focus on evaluating blood compatibility (*hemocompatibility*) because all materials used in cardiovascular applications come into contact with circulating blood. For this reason, several blood aspects must be addressed in all *in vitro* tests regarding the hemocompatibility (blood compatibility) of materials intended for cardiovascular applications.

Knowledge of what blood is, what the blood components are (as discussed in Chapter 2), and what events take place when blood interacts with a synthetic material surface, which is *not* the physiologic, normal, healthy, intact endothelium (as discussed in Chapter 5), is the most helpful background when evaluating the blood compatibility of cardiovascular biomaterials.

8.2.1 Hemocompatibility-Related Tests

Pertinent *in vitro* tests use established methods to determine activation and progression of the plasma coagulation cascade, for example, activation of Coagulation Factor XII (the surface factor) and production of thrombin. Several other tests examine various functions of activated platelets, including adhesion, aggregation, release of chemical compounds from their cytoplasmic compartments/storage pools, as well as synthesis and release of new chemical compounds (such as aggregating agents and growth factors). Formation of thrombin and activation of platelets when human blood comes in contact with the synthetic biomaterials tested are forewarning evidence of potential blood clot formation and subsequent serious problems when the material tested is implanted in the body. From the clinical perspective, blood clotting causes pathological complications which jeopardizes the welfare, and even threatens, the lives of patients.

Activation of the complement proteins should also be determined when evaluating the blood compatibility of biomaterials intended for cardiovascular applications because they instigate the activation of white blood cells (as discussed in Chapter 5). Red blood cells, the most numerous of the cell types in blood, may break down (a process known as "hemolysis") in the dynamic flow conditions of circulating blood when they forcefully come into contact with synthetic cardiovascular implants.

A unique complication related to cardiovascular applications is calcification. Calcium, an ion present in blood plasma, is a contributing component in several stages of the plasma coagulation cascade. In addition, calcium is a very important ion for the function of many organs in the body (i.e., heart, skeletal muscles, etc.). A calcium-related post-implantation complication occurred with heart valve implants containing mechanically moving leaflets composed of polymers. Calcium ions (from blood plasma) accumulated on the surface of those leaflets, causing calcification, making the polymer rigid (a potential source of material pieces breaking off and causing embolization), and affecting detrimentally the operation of the synthetic heart valve, specifically incomplete opening and closing in synchrony with blood flow requirements.

Other important aspects, which must be considered when evaluating the hemocompatibility of implant biomaterials, are the characteristics of the biomaterial surfaces because interactions of flowing blood in the cardiovascular system are exclusively surface-related interactions. Characterization of synthetic materials includes surface topography (such as smooth versus rough), chemical composition, electric charge (if that aspect is pertinent), hydrophobicity/hydrophilicity, and the type, concentration, distribution, and conformation of blood proteins adsorbed on material surfaces. The protein-related aspect is very important because (1) plasma proteins pre-adsorb onto material surfaces within seconds from the initial contact of whole blood with any synthetic material and (2) the adsorbed proteins subsequently dictate whether (and if so) which blood cells will interact with the surface under consideration.

8.2.2 Tests Under Static and Dynamic Conditions

Another unique aspect, which must be addressed in evaluating the hemocompatibility of implant materials, pertains to blood flow conditions. Continuously circulating blood inside the cardiovascular system is essential and vitally important for the welfare of humans and mammals. Uninterrupted blood flow must also be maintained when blood circulates within cardiovascular implants composed of synthetic materials. For this reason, the hemocompatibility-related tests must be conducted not only under static but also under dynamic (blood flow) conditions.

Many hemocompatibility-related tests (especially those adopted from hematology and physiology) have been performed under static conditions. In those cases, a specimen of the synthetic material tested is immersed in whole human blood *in vitro*, and it remains undisturbed for the duration of the test. Information obtained from tests under static conditions is valuable, but the hemocompatibility evaluation is not complete without results from tests under appropriate blood flow conditions.

Various hemocompatibility-related tests (discussed in Chapter 8, Section 8.2.1) can be conducted *in vitro* using specially designed and assembled laboratory setups simulating the *in vivo* dynamic flowing blood milieu. Only after successful completion of the *in vitro* tests under dynamic conditions, pertinent tests using experimental animals should be considered and, when approved by relevant regulatory committees, conducted (discussed in Chapter 8, Section 8.4). *Acute* animal tests are of very short duration but provide indispensable hemocompatibility-related information, which is a prerequisite for subsequent long-term animal tests to determine the pertinent performance of cardiovascular implants post implantation (discussed in Chapter 8, Section 8.4). At the end of each study, the animals are sacrificed, the implants excised and examined in order to determine pertinent aspects of biocompatibility, durability, function, and potential failure (discussed in Chapter 9).

8.3 Cultured Cell Models

Availability first, and then advances, regarding cultured, mammalian cells provided several pertinent *in vitro* models, which are now an integral part of the tests used in evaluating the biocompatibility of implant biomaterials and medical devices.

These techniques apply several lessons learned from biology (discussed in Chapter 4), specifically how cells interact with substrates. Briefly, cells do not interact with “bare” material surfaces; pre-adsorbed proteins are the mediators needed for the desirable outcomes, namely first adhesion and subsequent other mammalian cell functions and interactions with implant material surfaces. Another fact is that, within seconds from the introduction of materials inside the human body, they are covered by absorbed proteins (which are abundant in body liquids and tissues). These proteins mediate subsequent interactions of cells with substrates and promote cell functions (such as adhesion, migration, proliferation, synthesis, and deposition of extracellular matrix chemical compounds, etc.; discussed in Chapter 4), which are pertinent to new tissue formation – the most desirable outcome both in the process of wound healing and for the success of biomaterials and medical devices post implantation.

Additional reminders regarding the role of adsorbed proteins and subsequent cell interactions with biomaterials are the following facts (discussed in detail in Chapter 4): the biomaterial surface properties do affect and dictate the type, distribution, and conformation of absorbed proteins. The preadsorbed proteins dictate whether, cells will subsequently be recruited, whether they will adhere or not, and whether they will function or not. This knowledge is used in evaluating the biocompatibility of synthetic materials using cultured cell models.

8.3.1 Determination of Protein Interactions on Synthetic Materials

Studies using cultured cells in determining the biocompatibility of implantable materials should use appropriate tests to first examine interactions of proteins which are specific to the cells and tissues under consideration. Such select protein-related studies should be conducted and completed *before* any experiments involving cells.

The type and amount of each adsorbed protein can be determined using established laboratory techniques such as high-performance liquid chromatography (HPLC), which provides information (i.e., separation, identification, and quantification) for each dissolved component in a multicomponent solution. Proteins adsorbed on the surface of biomaterials can be stained with select staining markers and thus visualized using microscopy techniques. Availability of fluorescent stains and powerful microscopes enhanced visualization of pertinent details at the molecular level. An enzyme-linked immunosorbent assay (ELISA) allows detection and quantification of very small amounts of proteins. Based on molecular size, western blotting separates and identifies proteins in a mixture.

In addition to contributions to fundamental knowledge regarding proteins, investigation of the adsorption (including determination of the type, concentration, distribution, and conformation) of select proteins on synthetic materials provides insight regarding subsequent interactions of cells on those substrates.

8.3.2 Determination of Cultured Cell Interactions with Proteins Pre-adsorbed on Synthetic Materials

Successful completion of the experiments regarding protein absorption on biomaterial specimens is followed by tests using cultured cell models. For these studies, it is important to know anatomy, structure, and composition of the tissue which will be surrounding specific implants and then judiciously identify and choose the predominant cells. The chosen cells must either be excised from tissue specimens (using established techniques for cell harvesting) and appropriately characterized or be procured from appropriate, reputable, licensed vendors. Next, the cells are cultured in cell-culture laboratories and used in *in vitro* studies aiming at elucidating cell interactions with the biomaterials tested. Such studies provide cellular- and molecular-level information regarding how live cells from pertinent tissues (which eventually will be surrounding specific implants *in vivo*) interact with, and function on, the synthetic materials tested.

Since tissues contain more than one type of cells, once the aforementioned studies with the predominant cell type of a specific tissue are completed, these tests can be repeated using the other cell types (of the tissue) one at a time. Understanding the interactions of several constituent cells from a specific tissue with synthetic materials provides valuable insight into various aspects of the *in vivo* performance of the material under consideration and, thus, its suitability as substrate for the intended application.

There are many cell functions (discussed in Chapter 4) which can be evaluated by conducting experiments using cultured cell models. First and foremost is determination of cell survival and viability (i.e., cells must remain alive and functioning as they interact with the biomaterials tested). Cell damage or death is a most detrimental outcome, which disqualifies the material tested from further consideration for implant applications. Since cell viability is of utmost importance, the first assay to be conducted using cultured cells pertains to cytotoxicity. These assays examine whether contact of live, normal tissue cells with the new material or exposure to chemical compounds released from the new material under experimental *in vitro* conditions (which simulate those of the physiological environment, i.e., temperature, pH, etc.) either kill cells or affect detrimentally normal cell functions.

Once cell survival and viability are successfully established, then other cell-related tests can be conducted examining subsequent cell functions such as cell adhesion, morphology, migration, proliferation, occupation of the substrate surface, synthesis and deposition of extracellular matrix proteins, differentiation of stem cells, expression of programmed cell death (*apoptosis*), and, as needed, other functions pertinent to cells from specific tissues (as discussed in Chapter 4). Appropriate protocols for such assays have been developed, are available, can be obtained either from the scientific literature or from commercial vendors, and are used in biomedical research laboratories.

Other assays at the molecular and gene levels (e.g., quantitative polymerase chain reaction or qPCR and reverse transcription-polymerase chain reaction or RT-PCR, etc.) determine aspects at the RNA and DNA levels. Last, but not least, immunostaining assays identify the proteins produced by the cultured cells and deposited in their extracellular matrices on material substrates.

If contamination (for example, with bacteria) and, consequently, infection (as discussed in Chapter 6), are a concern regarding a specific application, separate experiments should be conducted using select types of pertinent pathogenic microorganisms. It should be emphasized that pathogenic microorganisms cannot be studied in the same laboratory with cultured mammalian cells because once bacteria and other pathogens contaminate either mammalian cell cultures or laboratories, it is very difficult to eliminate them. For these reasons, studies using various pathogenic microorganisms must always be conducted in separate, appropriately constructed, and maintained laboratories.

8.3.3 Cultured Cell Models: Closing Remarks

Overall, the cultured cell models are recognized as very effective screening methods for evaluating, in an expedited manner, many different types and a large number of new biomaterial candidates for implant applications. Moreover, these tests require a reasonable investment in laboratory equipment and facilities and modest operating costs; they are very good for studying functions, and pertinent mechanisms, of one cell type at a time.

Admittedly, the *in vitro* cultured cell models have advantages but also have certain limitations. These models provide the first valuable information regarding the performance of biomaterials in an environment that simulates several aspects of the biological-physiological milieu. The "traditional" static, cell-culture environment, however, does not completely duplicate (at best, it only simulates) the complex, multifaceted, interactive, and dynamic events which take place at the tissue-implant interface and around implants inside the mammalian body post implantation. For this reason, and whenever applicable, the *in vitro* tests using cultured cells should be conducted in bioreactors (which simulate dynamic mechanical force conditions) and in other laboratory setups (which duplicate the electromechanical milieu of specific tissues and organs).

When experiments using cultured cells are not successful, the evaluation process of the material under consideration stops. At that point, material scientists and engineers must start designing, formulating, and manufacturing new and different materials. In order to achieve subsequent success, information from the failed experiments (discussed in Chapter 9) must also be reviewed, and pertinent insight/knowledge should be utilized in improving current biomaterials. Most importantly, this information and knowledge should be incorporated into the design, formulation, and production of the advanced biomaterials of the future (discussed in Chapter 10). Recent research endeavors are using information and inspiration from biology and physiology and are pursuing "bioinspired" and "biomimetic" approaches to address limitations and complications associated with synthetic implant biomaterials. The focus is on providing synthetic materials with *biorecognition*, i.e., cues for proteins and/or recreating other appropriate signals for cells in order to promote biological/physiological processes leading to either production of new tissue or repair and replacement of damaged and/or diseased tissues in a precise, defined, and timely manner [2].

In closing, the pertinent realistic expectations from using cultured cell models are that they provide short-term, but important, biological-related information. That information, however, is limited, and it cannot be extrapolated into "global" conclusions regarding what may and may not happen when these materials are implanted inside the bodies of either animals or humans for long periods of time, specifically for the lifespan of the recipient patient. For this reason, additional and different types of tests are needed to complete the process of evaluating the biocompatibility of implantable materials and devices.

8.4 Animal Tests

Up to this point, the review of the process used in evaluating the biocompatibility of implants covered *in vitro* tests which were developed, and are used, in many traditional fields and laboratories (such as those in biomedical engineering, material science and engineering, biology, hematology, and biochemistry).

The next stage in the process of evaluating the biocompatibility of implants involves animal tests and is unique to biomedical engineering applications. The important prerequisite before any animal tests are even considered is that animals in research and testing should be used *only after* successful completion of all *in vitro* tests at the preceding stages (i.e., material and mechanical properties, cultured cell models, etc.).

Another unique aspect of this stage of evaluating the biocompatibility of implants is that use of experimental animals in research is regulated by federal laws in the United States. Biomedical engineers and scientists who are interested in implantable materials and devices should be aware not only of the existence but also of the important details of pertinent federal laws.

8.4.1 Regulations and Laws Regarding Animals in Research

The history of pertinent legal aspects started with the "Animal Welfare Act", which was signed in 1966; subsequent amendments were passed in 1970, 1976, 1985, 1990, 2002, 2007, and 2008. The Animal Welfare Act is enforced in the United States by the Animal Care Unit in the "Animal and Plant Health Inspection Service" of the Department of Agriculture [3]. Following pertinent investigations by the federal authorities and determination of violations of these laws may result in criminal penalties, civil penalties, and fines, as well as revoking permits to use animals in research and closing of laboratory facilities at universities, medical schools, companies, and other such institutions which conduct research using experimental animals.

The Animal Welfare Act of 1966 was the first to establish policies and standards regarding the handling, sale, and transport of animals, specifically cats, dogs, nonhuman primates, rabbits, hamsters, and guinea pigs. In addition, providers and dealers of experimental animals and the laboratories where these animals are used for research must be licensed and must provide identification (regarding the origin) of their animals. These requirements were included in order to prevent animal theft and exploitation; unfortunate examples are incidents of pets which, after they were stolen, were found in laboratories where research was conducted, and also when stray dogs and cats in towns and cities were collected by unscrupulous people and sold for research purposes. The Animal Welfare Act is the law prohibiting and punishing such practices [3].

Various amendments expanded the standards of the Animal Welfare Act of 1966 to include care of animals in commerce, exhibits, teaching, testing, and research. The list of protected animals was also expanded to include all warm-blooded laboratory animals such as cats, dogs, primates, and other mammals; birds, rats, and mice, however, were excluded from the protection provided by this law regarding the use of animals in research. Subsequent amendments further expanded and refined the original Animal Welfare Act; for example, the Amendment of 1985 addressed the welfare of animals in laboratories, the testing of animals, animals in exhibits, and captive marine mammals. The Congress of the United States established the Animal Welfare Information Center at the National Agricultural Library, a special service that provides information and resources regarding the care and welfare of animals as well as additional informational resources for alternatives [3]. Those alternatives are known as the *3Rs: Reduction, Refinement, and Replacement*, aiming at reducing the number of animals used for all types (including biomedical) research, refining techniques and appropriate tests, and developing novel and creative ways to replace the use of animals in research [4]. One recent pertinent development of major impact was the use of cultured cell models (discussed in Chapter 8, Section 8.3).

The major and foremost requirement of the Animal Welfare Act and of all associated amendments is the priority placed on the aspects that must be met at all times. This requirement is addressed and met by establishing special committees (for example, the Institutional Animal Care and Use Committee or IACUC) to oversee and, thus, assure proper animal care and welfare at institutions conducting research in the United States [3]. Researchers, who want to conduct experiments using animals at their home institutions, must first apply, and receive approval for the proposed protocols from the aforementioned committee (composed of faculty, staff [who have experience with animal-related research] and veterinary physicians who take care of animals) at their respective institutions. The submitted protocols must follow federal requirements for animal experiments. The submitted applications must include detailed description of animal handling at all stages, i.e., before experimentation, during, and at the end of the proposed study. Experiments start only after the home institute IACUC approves the proposed protocols. Preapproval of animal-related protocols by the IACUC at the respective home institute is required before consideration of an animal-research-related grant application submitted to the National Institutes of Health, the national agency of the United States which funds biomedical research [3].

Biomedical engineers who are considering animal tests in order to determine the biocompatibility of their biomaterials and implants should be aware of the following specific requirements. The federal animal research regulations in the United States identify and specify the protected animal species; for example, pets are protected, but farm animals are not. How animals are procured for research purposes is also addressed; for example, using stray cats and dogs is against animal procurement laws. In fact, animals used for research are raised on special farms which are licensed vendors and have established professional credentials [3]. Stating credible sources and vendors regarding the animals which researchers plan to use in their studies, provides assurance to reviewing committees of funding agencies (such as the National Institutes of Health) that the proposed study meets the animal procurement requirements.

How animals will be bred or grown is another specific requirement regarding animal tests. Some studies, investigating molecular- and gene-level aspects of diseases, are long-term endeavors requiring several generations of, for example, mice. In addition to procurement of mice from licensed vendors, the animals in such studies must be bred for several generations at the location where the study is conducted. For these reasons, the proposed research protocols must address all pertinent aspects in order to obtain approval from the appropriate committees and conduct the proposed study. Information regarding veterinary care of animals upon arrival, before, during, and at the end of every study is required and must be provided by veterinary doctors and appropriately trained personnel who take care of the animals for the duration of the study [3].

An institution where animal tests are conducted must also have an appropriate animal facility. The Animal Welfare Act and subsequent amendments specify and define regulations pertinent to animal facilities [3]. These details include continuous maintenance of cleanliness, temperature control at all times, durations of lights on and off (to control the biological rhythm of animals because they are continuously kept indoors), the number of animals per cage, the dimensions of animal cages in accordance with the size of the housed animals, and availability of water and appropriate food for the animals. Food should not be given to the animals prior to surgery; water, however, must be available to them at all times.

Special animal requirements are also addressed by the laws; for example, the animal facility which houses dogs should provide a room or space (i.e., runners) where the dogs are allowed to run and exercise several times a day. Use of primates in research is very strictly controlled; even the psychological welfare of these animals must be taken care of.

In addition to having animal facilities, institutions (where animals are used in research) must also provide information regarding who is in charge of the animal facility, including the training and pertinent experience of animal care personnel. The professional credentials (i.e., training, expertise, veterinary practice-related licenses, etc.) of veterinary physicians, animal surgeons, and assistants, as well as the credentials and experience of the researchers (who conduct the animal studies) must also be reviewed and approved by the appropriate regulatory committees.

Applications seeking approval of experimental protocols using animals are lengthy documents and must include the following information: detailed description of all experimental procedures which include animals; pertinent controls at all stages of the study; detailed surgical procedures; anesthesia used during surgery as well as other tranquilizing drugs; details regarding pre-, during, and post-surgery animal care; the credentials and accreditation of the personnel working in the animal facility; type, dose, and duration of delivery of prescribed medications; anticipated potentially adverse effects of medical treatments and details regarding medication type and dose; euthanasia (type and administration) at the end of the study; information regarding disposal of dead animals either during or at the end of the study; and the medical treatment of animals when they are sick for reasons unrelated to the study.

If animals during a study get sick, and that development is related to the study in progress (e.g., involving implants), that result must be included in appropriate documentation, collected data, and pertinent reports. All adverse incidents should be considered when the outcome of the study is evaluated. If, however, animals get sick for reasons unrelated to the study, the sick animals must be treated by the veterinary physician; pertinent information should be documented in the records of the study and, if the animals recover, they are kept in the study.

In addition to detailed and lengthy applications, which must be submitted to appropriate regulatory committees (in order to obtain approval to conduct experiments using animals) before the project starts, lengthy, very detailed, and meticulous record-keeping of all pertinent aspects of the study is required as the study progresses. These aspects include detailed description of all experimental procedures involving animals, daily logs, descriptions of the respective controls, and comparison of those results with the results obtained from the animal experiments, details of any additional surgery procedures, pre- and post-surgery treatment of the animals, and statistical analyses of the data. All aforementioned aspects must be recorded for the duration of each study. Such records must be kept because reports are required by the regulatory committees (such as the Institutional Animal Care and Use Committee) and national funding agencies (such as the National Institutes of Health) at specific and regular time intervals during, and at the end of, the study. Last but not least, all aforementioned documentation must be collected, maintained, and submitted to the FDA when researchers (who conducted those studies) and the companies (which produce the products under consideration) seek FDA approval and, thus, permission to market legally approved products that help patients.

8.4.2 Special Considerations Regarding Animals Used in Research

Use of animals in experiments must meet unique and special requirements, which are not applicable in traditional engineering disciplines. Biomaterial biomedical engineers, however, must be aware that all animals in experimental studies should be treated with *humane care*. Specifically, all animal-involving tests and experiments should not expose animals to unnecessary pain and suffering [5]. Studies using animals should use only the numbers of animals needed to reach statistically significant results of the collected data. Unnecessary duplication of studies should be avoided by first searching previous similar studies that had been conducted at a specific laboratory at a university, medical school, and company as well as by thorough search and review of the scientific/engineering literature. Such practices will avert unnecessary duplication of experiments and avoidable use of large numbers of animals in research [5].

Concerns regarding the use of animals in research have prompted search for alternatives. This approach is exemplified by the "3Rs: Reduction, Refinement, and Replacement" (discussed in Chapter 8, Section 8.4.1) [4]. An interesting and promising approach is the establishment of the "Medical Device Development Tools" program in the Food and Drug Administration Center for Devices and Radiological Health. The purpose of this program is to explore and establish replacements for animal tests. One such development is using cultured cells (discussed in Chapter 8, Section 8.3) as part of the *in vitro* evaluation of the biocompatibility of materials and devices. Another very promising development is the "organ-on-chip" model. These are sophisticated, *in vitro* laboratory systems which can be used to determine the biocompatibility of novel materials by exposing them simultaneously to cells from several pertinent tissues as well as to various body liquids simulating the *in vivo* milieu of the human body.

In addition to knowing the special aspects pertinent to research using animals, biomedical engineers must also be aware of relevant compliance requirements. Individual investigators, groups of investigators, and institutions (where animal research is conducted) must comply with all local, state, and federal laws and regulations regarding animal tests.

The federal laws established in the United States (starting with the Animal Welfare Act of 1966 and subsequent amendments) have addressed various aspects regarding the care and use of laboratory animals and established federal laws and regulations for monitoring and surveilling animal-related research. Compliance to all aforementioned regulations and laws is required by the federal agencies which fund research projects (for example, the National Institutes of Health in the United States). Moreover, compliance is scrutinized by local review committees (i.e., IACUC), which approve animal-related research at universities, medical schools, and industry.

8.4.3 Nonfunctional, *Ex Vivo*, and Functional Tests

The biocompatibility of biomaterials and medical devices using animal studies is examined and determined in stages using different experimental procedures.

The first such tests are *nonfunctional*. In this case, a small sample of the candidate new material is placed under the skin on the back of experimental animals (usually rats or mice). Implantation of these samples involves a limited surgical procedure and requires small cuts of the skin (before placing the material sample under the skin of the animal), and a few sutures at the beginning and at the end of the surgical procedure. After implantation, the animal response is monitored for evidence (and progress) of inflammation (specifically, for the cardinal signs of local redness, swelling, and increased temperature (discussed in Chapter 2, Section 2.5); for this reason, animals with light skin are preferred and used in such tests. Development of chronic and systemic inflammation or implant rejection, however, provide evidence that the tested material is not compatible and should not be considered for implants. If there are no adverse reactions, the study is completed. Excision and removal of the tested material specimens require limited surgical procedures.

Ex vivo are short-term experiments that have been used for evaluating the blood compatibility of vascular grafts [6]. In this case, a candidate vascular graft (tubular specimen composed of a new polymeric material) is formed into a U-tube (fistula). Each end of this U-tube is introduced in the exposed artery and vein of the leg of an experimental animal (under anesthesia) and used as an arteriovenous shunt. Blood from the animal circulation then flows through the closed U-tube channel (i.e., from the animal, through the U-tube, and back to the animal) for a predetermined period of time (in the order of minutes to few hours). The U-tube/fistula is then carefully detached, the cut veins and skin are sutured, and the animal recovers from anesthesia. The interior of the U-tube is then examined for the presence of blood clots (for size, distribution, and type) as a function of time of exposure to flowing blood (discussed in Chapter 8, Section 8.2.2). The collected data provide the evidence needed to reach appropriate conclusions regarding blood compatibility (hemocompatibility) of the material tested under blood flow conditions (discussed in Chapter 8, Section 8.3).

After successful completion of the nonfunctional and *ex vivo* tests, *functional tests* are considered and conducted. Functional tests are most important because, in this case, both the constituent materials and a prototype of the medical device are tested and evaluated inside the body of animals for long periods of time. For functional tests, a prototype of the specific implant (which is eventually meant for human use) is designed, scaled down, and formulated for the size of the recipient animal. For example, in the case of hip prostheses, biomedical engineers design and produce a prototype hip prosthesis for rabbits (the preferred animals for these studies). Surgeons replace one natural hip of an animal with the model hip prosthesis implant. The pertinent surgical procedure is extensive and involves total anesthesia of the animal. The post-implantation time period of such study is comparable to the expected duration for the presence of the specific implant in the bodies of recipient human patients. Post implantation and during long-term studies, the recipient animals are taken care of by specially trained animal care personnel and veterinary doctors. Progress of the experiments is monitored by the researchers who conduct the study.

If the implant is a vascular graft, the pertinent surgical procedure first removes part of a blood vessel before replacing it with a properly designed, sized-down, and formulated vascular graft composed of the new synthetic polymer under consideration [7]. At the end of the experiment, the synthetic vascular graft is excised, and its interior is examined for the presence of blood clots, including size, distribution, and type (discussed in Chapter 3). The collected data provide the evidence needed regarding the blood compatibility (hemocompatibility) of the material tested under blood flow conditions (discussed in Chapter 8, Section 8.2).

In all studies, it is important to monitor, follow up, and record not only implant functionality and local reactions but also any adverse systemic reactions which may affect the welfare of the recipient animals.

The functional tests require extensive surgical procedures, recovery of the animals from anesthesia and surgery, and follow-up for long periods of time in order to determine not only the biocompatibility of the biomaterials and of the medical devices tested but also the functionality of the prostheses. In the case of hip prosthesis, pertinent information is obtained from the performance of the implant under mechanical loading when, for example, the recipient animal walks, etc., post implantation; in the case of vascular grafts, either minimal blood clotting or absence of blood clots and of blood circulation problems are the desirable outcomes.

In summary, the knowledge provided by the successes and failures of all aforementioned *in vivo* animal tests and studies is needed in order to determine whether the biomaterials and the implants under consideration can be used in the next stage of their biocompatibility evaluation, which is implantation inside the human body. It should be mentioned, however, that successful performance of implants inside animals does not always guarantee success of those models in humans during Clinical Trials.

8.4.4 Closing Remarks

In closing, animal tests are serious and major undertakings, which must be considered only when all the preceding biocompatibility-related *in vitro* tests have been successfully completed. The studies which use animals are expensive because they require special facilities, procurement, and maintenance of animals on a daily basis (because animal facilities charge per diem for lodging and food), as well as for care of the animals by properly trained personnel and by veterinary doctors (who monitor the health and treat the animals when they get sick).

Experiments using animals are complex but invaluable because they provide information regarding the performance of biomaterials and implants in the bioactive, multifaceted, and dynamic biological milieu. The results of animal tests are difficult to interpret: certainly, negative results make the specific biomaterial and medical device unacceptable for further consideration. In addition, negative outcomes confirm and establish the unacceptability of the tested biomaterial and implant for humans. Positive results, however, do not necessarily prove or predict the biocompatibility of those implants in humans. Overall, animal experiments are useful and unavoidable precedents for clinical trials when new materials and medical devices are implanted in humans.

8.5 Clinical Trials

A characteristic of the progress made along the stages of evaluating the compatibility of implantable biomaterials and devices is that, once a stage (e.g., the *in vitro* material mechanical properties) is successfully completed, the researchers involved in the project can decide to move onto the next stage (specifically, the *in vitro* tests using cultured cell models). Continuing onto animal tests, the next stage, however, requires not only the decision of the researchers involved in the project but also appropriate approval and, thus, permission from pertinent regulatory committees and agencies at the local, state, etc., levels in order to proceed with the animal tests.

Even when all preceding stages, that is, all the *in vitro* and then the animal tests, are completed successfully, the decision to proceed to clinical trials, the next stage of the biocompatibility evaluation process, cannot be made *only* by the health-care personnel, the researchers (who developed the product under consideration), and the companies (which want to produce and market the new implant). That decision requires approval and permission by regulatory agencies such as the FDA in the United States and (if the implant will be available in the global markets) by similar agencies in other countries around the world.

In order to complete the process of evaluating the biocompatibility of implants, the last but most important stage, is clinical trials. At this stage, new products are tested and evaluated following implantation in human volunteers in studies conducted under close, and stage-by-stage, surveillance by the pertinent regulatory agency (e.g., the FDA). Successful completion of clinical trials under the aforementioned conditions is needed in order to receive permission for legal marketing of pertinent products and use of new implants in patients around the world. This is, of course, the *ultimate objective* for all biomedical implants.

8.5.1 An Introduction to the Food and Drug Administration (FDA) of the United States (US)

The FDA is a regulatory government agency, which was established following the "Pure Food and Drug Act of 1906". This law was passed by the US Congress and became the first consumer protection law in that country. Subsequent amendments expanded and strengthened the 1906 Law.

Mandated by the US Congress, the keywords in the FDA mission statement are "safety" and "efficacy" [8]. These aspects are of paramount importance, are given priority, and are required for all devices implanted into human patients. At the same time, the FDA is striving to assure safety and efficacy without delaying and preventing innovation and availability of new medical products and devices.

In complying with its mission, the FDA reviews and assesses applications proposing evaluation of various products (including food, drugs, cosmetics, animal foods, dietary supplements, medical devices, biological goods, biological products, and biomaterials), monitors pertinent evaluation processes, and approves successful endeavors [8]. As a regulatory agency, the FDA oversees the safety and efficacy of drugs and medical devices before such products are approved for marketing in the United States [8].

In addition to complying with FDA regulations and obtaining approval by that agency, if a product will be sold in global markets, the requirements of regulatory agencies in other countries around the world must be observed, and pertinent additional approvals must be obtained. Examples of such regulatory agencies are the International Organization for Standardization, the Organization of Economic Co-operation and Development, the Japanese Ministry of Health, Labour and Welfare, etc.

8.5.2 Requirements for Clinical Trial Applications

Detailed information regarding the requirements for clinical trials is available in publications by regulatory government agencies (e.g., the FDA in the United States and by similar regulatory agencies in other countries around the world) as well as by various professional organizations (e.g., American Cancer Society, Medical Device Diagnostic Industry, etc.). These sources have compiled, and regularly update, detailed instructions regarding applying and obtaining approval for initiating, conducting, and completing clinical trials [9].

Requests for clinical trials in the United States must first be submitted to, and approved by, the FDA [10]. These applications include the following information: the results of successful completion of all preceding biocompatibility evaluation stages (i.e., all *in vitro* and animal studies); detailed information regarding the device (such as material content, assembly, manufacturing, etc.); detailed protocols describing the proposed clinical trials; information (i.e., medical training, credentials, expertise, etc.) of the participating health-care personnel (i.e., doctors, surgeons, nurses, etc.) and the credentials of the animal care personnel; and information about the hospital(s) where the proposed clinical trials will be conducted.

The FDA in the United States (and similar government regulatory agencies in other countries) require that applications for clinical trials (which are submitted to them) include all records and results from every preceding stage of the biocompatibility evaluation process pertinent to the new implant under consideration [10]. In addition to meticulous, detailed, and complete records of the procedures used and of the results obtained from the *in vitro* materials and mechanical tests, *in vitro* cellular models, and animal tests, these applications should include descriptions of the new device, surgical procedures, pre- and post-operative and medical care of patients, use of appropriate controls, and evidence that appropriate statistical analyses will be applied in determining the statistical significance of the results regarding the new product. All aforementioned documents *must* be submitted to the FDA in the United States (and, as needed, to appropriate government regulatory agencies of other countries) as part of submitted applications requesting approval to conduct clinical trials for new implants [10].

8.5.3 Patient Participation in Clinical Trials

Following approval by a pertinent government regulatory agency, all clinical trials have a unique and most important feature: human volunteers participate in these trials. For this reason, clinical trials have special requirements.

Ethical considerations pertain to the selection of patients who will participate in a clinical trial because *only* the patients chosen for the study have a chance to benefit from the new medical device. These consequential decisions regarding patient selection are made by the medical doctors and the health-care providers in charge of Clinical Trials.

Participation of humans in Clinical Trials warrants important and special considerations. The rights and dignity of the participating human subjects should always be protected. Participating patients must give their *informed consent* before they participate in a clinical trial [9, 10]. The identity and all details on the medical records of participating patients must remain confidential *forever*. This is the reason why (when, for example, the results of clinical trials are reported) neither names nor other identification details of the participant patients are revealed; in fact, each participating patient is given a number, by which they are referred to in all pertinent reports.

8.5.4 Phases of Clinical Trials

In the United States, the various stages of clinical trials are known as *Phases* followed by Roman numerals and designated as Phase I, Phase II, Phase III, and Phase IV. In practice, the aforementioned Phases of the clinical trials must be followed sequentially. The number of participating patients, duration of the studies, and the number of hospitals (where Phases I, II, and III are conducted) increase with each Phase. Phase IV trials evaluate the safety and efficacy of approved implants over long periods of time and involve large numbers of patients [9, 10]. The results from each completed Phase are reported to the pertinent government regulatory agency (e.g., FDA) for review and approval *before* continuing onto the next Phase of the pertinent clinical trial.

Very special cases of clinical trials, which precede even Phase I, are those which require "Emergency Use Authorization" [11]. The rationale for such requests are for humanitarian reasons. Attending physicians, for example, may apply to the FDA for such authorization in order to use a new product or device before its biocompatibility evaluation has been completed or approved yet. In this case, the attending physicians must prove that they have exhausted all other available treatment options, the patient is very sick, and the prognosis for patient survival is either extremely low or none. When the requested FDA authorization is given, the new product or device is used only in one patient. The outcome of such intervention is that, although the volunteer patient's condition may improve for a short time (i.e., for a few days or weeks), death is inevitable. Invaluable information, however, is obtained regarding the performance of new devices inside the human body.

An example of an emergency use authorization case was the first time an artificial heart (composed of synthetic materials) was implanted to replace the malfunctioning heart of a very sick patient in a South Africa hospital in the 1970s [11]. That was the first ever treatment using a device composed of synthetic materials to replace the human heart. The implanted artificial organ extended survival of the recipient, very sick patient, for a short period of time. The *in vivo* information this case provided to attending physicians, surgeons, and the researchers (who designed and fabricated the artificial organ) was invaluable and most instructive for subsequent research endeavors for new and improved cardiovascular implants [11].

8.5.5 Requirements for Successful Clinical Trials

In terms of requirements, implant biomaterials and medical devices (which are composed of those biomaterials) must *always* be safe for both the recipient patients and the health-care providers and must comply with regulations of various government regulatory agencies.

In order to obtain approval by the FDA in the United States, the results of the clinical trial evaluation of a new device must meet safety and efficacy requirements: specifically that there are neither risks nor complications which may jeopardize the welfare, health, and life of the receiving patients. Furthermore, in comparison to other previous or existing similar implants, the new product must provide additional and improved benefits. Successful completion of the clinical trials is required in order to obtain approval from the government regulatory agencies and, thus, bring new biomedical devices to legal medical markets and to the patients who need them and who will benefit from the availability of the new products.

8.5.6 Undesirable Outcomes

Unfortunately, some new devices fail during clinical trials, and thus, they do not result in marketable products. In such cases, the clinical evaluation process is stopped, and the new product is removed from any further consideration. Subsequent examination and evaluation of excised failed implants provide information which biomedical engineers and scientists must use in designing, formulating, and producing the improved biomaterials and devices of the future (discussed in Chapter 9).

In terms of consequences, failure of implants post implantation associated with injury and/or death of recipient patients may lead to litigation. The company producing the failed implants, as well as the individual scientists and engineers involved in that project, may be brought to court in *class-action lawsuits* pursued by lawyers representing either patients (who suffered because of failed implants) or members of the families of patients (who died because of implant failures). Such events are reported in the popular press and attract a lot of attention from the general public. When, in the past, cases of *in vivo* failure of new biomedical products were brought to jury trials in the United States, the patients and their lawyers won many cases and received monetary settlements.

8.5.7 Closing Remarks

Success of all *in vitro*, cultured cell models, and animal tests stages of the biocompatibility evaluation process, provides valuable information, *but* it does *not guarantee success of implants in humans*. Because human volunteers are involved, clinical trials are initiated only after they receive approval from appropriate government regulatory agencies and are conducted under the surveillance of such agencies. Successful clinical trials are needed to establish *safety* and *efficacy* of every new devices under consideration for subsequent clinical use. Only then new implants receive approval from appropriate government regulatory agencies and become commercially available for the patients who need them.

Clinical trials are serious, multifaceted, and very demanding medical and scientific endeavors. In addition to scientific aspects, these studies require major investments of time and money. All pertinent special and extra requirements, however, are justified because successful completion of clinical trials establishes the ultimate biocompatibility of implants in the human body, the milieu for which they were designed.

References

1 Agrawal CM, Ong JL, Appleford MR, Mani G. *Introduction to Biomaterials: Basic Theory with Engineering Applications*. Saltzman WM, Chien S editors. Cambridge University Press; 2014.

2 Lumelsky NL. Commentary: engineering of tissue healing and regeneration. *Tissue Eng.* [Internet]. 2007 [cited 2023 Aug 31];13(7):1393–8. Available from: https://pubmed.ncbi.nlm.nih.gov/17550337/.

3 Cardon AD, Bailey MR, Bennett BT. The Animal Welfare Act: from enactment to enforcement. *J. Am. Assoc. Lab. Anim. Sci.* [Internet]. 2012 [cited 2023 Aug 31];51(3):301. Available from: https://www.ncbi.nlm.nih.gov/pmc/articles/PMC3358977/.

4 Hubrecht RC, Carter E. The 3Rs and humane experimental technique: implementing change. *Animals* [Internet]. 2019 [cited 2023 Aug 31];9(10). Available from: https://www.ncbi.nlm.nih.gov/pmc/articles/PMC6826930/.

5 Stokes WS. Humane endpoints for laboratory animals used in regulatory testing. *ILAR J.* [Internet]. 2002 [cited 2023 Aug 31];43(Suppl 1):S31–8. Available from: https://dx.doi.org/10.1093/ilar.43.Suppl_1.S31.

6 Lelah MD, Lambrecht LK, Cooper SL. A canine ex vivo series shunt for evaluating thrombus deposition on polymer surfaces. *J. Biomed. Mater. Res.* [Internet]. 1984 [cited 2023 Aug 31];18(5):475–96. Available from: https://onlinelibrary.wiley.com/doi/full/10.1002/jbm.820180503.

7 Weber M, Steinle H, Golombek S, Hann L, Schlensak C, Wendel HP, et al. Blood-contacting biomaterials: in vitro evaluation of the hemocompatibility. *Front. Bioeng. Biotechnol.* [Internet]. 2018 [cited 2023 Aug 31];6:99. Available from: https://www.ncbi.nlm.nih.gov/pmc/articles/PMC6054932/.

8 Fleming TR, Demets DL, McShane LM. Discussion: The role, position, and function of the FDA—the past, present, and future. *Biostatistics* [Internet]. 2017 [cited 2023 Aug 31];18(3):417. Available from: https://www.ncbi.nlm.nih.gov/pmc/articles/PMC5862299/.

9 Society AC. Clinical trials: what you need to know [cited 2023 Aug 31]; Available from: https://www.cancer.org/content/dam/CRC/PDF/Public/6800.00.pdf.

10 Conducting Clinical Trials | FDA [Internet]. [cited 2023 Aug 31]. Available from: https://www.fda.gov/drugs/development-approval-process-drugs/conducting-clinical-trials.

11 Khan S, Jehangir W. Evolution of artificial hearts: an overview and history. *Cardiol. Res.* [Internet]. 2014 [cited 2023 Aug 31];5(5):121. Available from: https://www.ncbi.nlm.nih.gov/pmc/articles/PMC5358116/.

9

Lessons Learned from Implant Failure, Retrieval, and Evaluation

Before implantable biomaterials and medical devices designed, formulated, assembled, and produced by biomedical engineers are used to help patients, the biocompatibility of all such new products must be properly evaluated in order to identify, and eliminate, any aspects which may cause problems post implantation. Moreover, obtaining approval from pertinent regulatory agencies (such as the Food and Drug Administration (FDA) in the United States and similar agencies in other countries) provides the required and important assurance regarding the safety and efficacy of implantable devices and appropriate authorization before such implants become available in national and international markets and are allowed to be used in patients. Despite all good intentions and appropriate preparation, however, some implant biomaterials and medical devices fail post implantation.

9.1 Complications Leading to Implant Failure Post Implantation

Information regarding failure of various implants can be found in medical literature reports. Some reasons of failure (e.g., infection) are prevalent and affect all types of implants independent of the implantation anatomical site. Other aspects are specific to the organ and part of the body for which each synthetic implant was designed and used. All such complications are undesirable clinical outcomes.

Infection (discussed in Chapter 6) is always a possible major problem and anathema for all implantable devices when pathogenic microorganisms attach and grow in colonies either on or in the vicinity of implants; this situation becomes more serious when pathogens form and protect themselves under biofilms (discussed in Chapter 6: Sections 6.1.1, 6.1.2,

Concepts of Tissue-Biomaterial Interactions: Fundamentals and New Directions,
First Edition. Rena Bizios and Marissa E. Wechsler.

Companion Website: www.wiley.com/go/Concepts_of_Tissue-Biomaterial_Interaction_1e

and 6.2.2). Pathogenic microorganisms inside biofilms are very difficult (if not impossible) to annihilate by treating patients with medications (such as antibiotics discussed in Chapter 6, Section 6.2.4). Most importantly, biofilms protect pathogens from the natural defense mechanisms of the human body, i.e., the white blood cells, which are designed by nature to detect and kill invader microorganisms. For these reasons, and as pathogens thrive, the resulting infection may increase locally, and it may spread systemically to other parts of the body.

Many times, healing (the desirable outcome of the surgical procedures which place implants inside the bodies of patients) is either not complete or it is insufficient, or it may cause excessive, undesirable (and even pathogenic, e.g., cancer) tissue growth (discussed in Chapter 3: Sections 3.6 and 3.8). Other adverse local and systemic tissue reactions include inflammation and toxicity (which kills cells); both events affect detrimentally surrounding tissues and their function.

Fragments of synthetic materials from implants (such as polymers, metals, etc.), may break off or may be produced from unwanted dissolution of synthetic materials. When they are present in blood or in the lymphatic system, these by-products may be transported to other parts of the body and cause health-related problems.

Hydrolysis involves infiltration of water (of which there is plenty in the liquids, tissues, and organs of the human body). When water infiltrates the bulk of implanted polymers, it may cause several problems: materials may swell, the size and shape of the implant may change, the carbon backbone of polymers may break, and metals may corrode. Ensuing changes in the characteristic chemistry, shape, and geometry may also cause problems with the mechanical properties of implants.

In all cases, when, as a result of the aforementioned complications, implants fail post implantation, regulatory agencies (such as the FDA in the United States) recall their approval; in addition, such implants may be excised surgically and be removed from the market.

9.2 Examples of Vascular Implant Failure Post Implantation

Blood clotting (discussed in Chapter 2: Section 2.3) causes major and critical complications for cardiovascular implants. Examples include complete occlusion of vascular grafts and embolism (a disastrous occurrence when a blood clot from one part of the body is lodged in a blood vessel of smaller diameter and stops blood flow). When blood does not clot (as in the case of hemorrhage), complications arise because of blood loss, which the human body needs for survival and good health. Hemorrhage may result from prescribed medications, which patients receive after implantation of cardiovascular devices in order to prevent blood clot formation when the patient's blood comes in contact with the synthetic material of cardiovascular implants.

In the case of vascular graft failure, the first and foremost reason is formation of various sizes of blood clots: some of them are big enough to occlude completely the lumen of vascular grafts and others, dislodged from the site of their original formation and, as emboli in the dynamic milieu of blood, travel downstream and occlude another blood vessel of smaller internal diameter.

Infection, always a big problem with all implants including cardiovascular ones, must be avoided at all costs.

Aneurism, an unexpected large expansion of the natural vascular tissue (connected at the two ends of the vascular graft), sometimes may reach the limit of the mechanical properties of the natural tissue and burst; as a result, blood spills out of the vasculature, breaking the closed blood circulatory system and endangering patients' lives.

Anastomotic hyperplasia is the case of excessive growth of natural vascular tissue at the points where the natural tissue is sutured with a vascular graft. This excessive tissue growth is the result of proliferation of smooth muscle cells from the vascular tissue and excessive deposition of their extracellular matrix. This new tissue mass grows toward the center of the vascular graft, causing stenosis or narrowing of the path through which blood must flow. Potential undesirable results in such cases are blood clots and much harder pumping by the heart in order to deliver adequate blood flow through the anastomotic hyperplasia stenoses.

Disintegration or degradation is not part of the design, formulation, and purpose of vascular grafts; if, and when, it may happen, however, it is a major unwanted complication of detrimental clinical importance.

Another chemical- biochemical-related reaction is mineralization/calcification. Several ions are present in blood, other body liquids, and human tissues. Calcium, for example, is a very important ion for the proper function of the heart. When, however, calcium ions concentrate and accumulate into implantable devices (especially on parts where they are not supposed to be), they may cause major material failure and malfunction of implants.

An illustrative example was reported in the medical literature regarding certain heart valves. In imitation of the natural heart valves, those heart valve models (composed of synthetic materials) had leaflets which opened when blood passed through and then closed. Following appropriate procedures, this new heart valve received approval by regulatory agencies and was implanted in humans. Clinical complications arose because of related malfunction. Examination of failed specimens revealed that calcium (present in the blood plasma) had accumulated preferentially on the moving leaflets of the synthetic implants, rendering them rigid, and hindering their intended complete opening and closing. The failed model was removed from the medical market, and its production in industry was stopped.

Physicochemical conditions also may cause failure of biomaterials as a result of proteins, water, and lipids (from physiological liquids and tissues) adsorbing onto the surfaces and, sometimes, infiltrating the bulk of implanted materials. Adsorbed proteins and lipids can affect not only the appearance (such as color) but also the chemistry and the mechanical material properties of implants. Monomers from polymers may leach out, accumulate in the vicinity of implants, and cause unwanted local inflammation and/or toxicity. A pertinent example is one from the fabrication process of polymers. In order to improve the fluidity of polymers (and thus facilitate molding), nonbiologic, volatile chemical compounds (known as "plasticizers") are used. When plasticizer molecules remain on implantable materials (e.g., because they do not evaporate when plastics solidify during manufacturing), they leach out *in vivo* post implantation and cause problems in the surrounding tissues (such as toxicity, inflammation, and related pathological complications).

9.3 Reasons for Failure of Orthopedic and Dental Implants Post Implantation

The list of reasons responsible for the reported failed orthopedic and dental implants is long and pertains to the dynamic conditions under which such implants exist and continuously perform *in vivo*.

Creep, wear, stress propagation, and fracture are mechanical-related failures from either exposure of materials to dynamic mechanical environments or during mechanical performance of devices. Because all these problems are undesirable, the design and *in vitro* testing of implants should address potential issues ahead of time and eliminate them before the device is implanted either in animals or in humans.

When implantable materials are not designed on purpose to degrade *in vivo*, possible, relevant, and undesirable results of the degradation by-products, both locally and systemically, should be addressed and resolved ahead of time during the design and *in vitro* testing of those materials in order to avoid medical complications post implantation.

A notable exception, specifically the case of desirable degradation of material scaffolds for tissue engineering applications, was discussed in Chapter 5: Section 5.6. For that special application, degradation is incorporated intentionally in the synthetic material; in fact, those materials are designed to be dissolved in synchrony with new tissue formation, which then replaces the synthetic material. Degraded material by-products are excreted through natural processes of the body. In contrast, unexpected and unpredictable implant material degradation post implantation may lead to complete failure of functioning implants, as well as to clinical complications threatening the welfare, and sometimes the lives, of patients.

Chemical and biochemical (e.g., oxidation/reduction) reactions can contribute to malfunction and even failure of biomaterials and medical devices post implantation. Long-term exposure of metal implants in the water environment of the body may cause corrosion; the type, concentration, and bioactivity of released and accumulated ions may result in physicochemical events with detrimental effects locally and systemically.

9.4 Examples of Orthopedic and Dental Implant Failure Post Implantation

The medical literature contains reports of major orthopedic and dental (bone) implant failure post implantation which is either incomplete or lacking integration of these implants into the surrounding bone.

Incorporation of orthopedic implants (such as hip prostheses) into surrounding bone is needed in order to immobilize the prosthesis *in situ* for proper function under the dynamic loadings experienced during use of those implants by patients post implantation. Loose implants in bone translate into major trouble for the implant and for the patient.

Sometimes incomplete integration is a result of bone resorption. For example, bone resorption in hip prostheses results from uneven distribution of stresses when these prostheses are loaded post implantation as patients walk, climb stairs, etc. In this case, the mechanical loading is preferentially distributed through the metal (because of its stronger mechanical material properties) and not through the bone [1]. It should be noted that bone is a tissue that needs to be exposed to mechanical stimuli in order to maintain its viability, structure, and function.

Other reasons for orthopedic implant failures are due to fixation failure [1]. When surgeons implant a total hip prosthesis, they need to form space in the bone marrow of the femur to accommodate the stem of the hip prosthesis; for that purpose, they remove a large amount of bone marrow tissue. The stem of the prosthesis is then placed in that site, surrounded by a gap in between the implant metal and the femur bone. In order to stabilize the prosthesis in situ, that surrounding gap is filled with polymethyl methacrylate (colloquially known as "bone cement"), a thermoplastic resin which is mixed, prepared at the patient's side, and used immediately. Polymethyl methacrylate polymerizes into a rigid plastic, fills the gap, and, thus, stabilizes the hip prosthesis in situ.

Interface separation is another very specific reason for failure of bone implants [1]. Success of bone implants (mostly metals) requires formation of new bone at the tissue-implant interface; for this purpose, pertinent functions of osteoblasts (the bone-forming cells) are required. In order to improve osteoblast functions relevant to new tissue formation, metal implants were coated with ceramics (specifically hydroxyapatite for orthopedic, and alumina for dental, implants) because osteoblasts (the bone forming cells) interact favorably with ceramics. The result was enhanced interaction of osteoblasts as well as strong bonding of the newly formed bone on the ceramics. The reason for related implant failure in such cases was delamination (i.e., separation into constituent layers) at the interface between the ceramic coating and the metal surface (and not between the bone and ceramic layer). This outcome, known as "interface separation loosening", is unquestionable evidence of the bonding strength of bone on ceramics [1].

Interface separation also results from fixation failure, either lack of or limited new bone formation around implants, and possibly from bone resorption by osteoclasts (the bone-resorbing cells). In all aforementioned cases, the undesirable outcome is loosening and subsequent relative motion in the dynamic milieu of orthopedic/dental implants; such developments cause malfunction of implants and coincide with complaints from inflicted patients regarding pain related to the unsatisfactory performance of those implants. Such outcomes are major contributors to orthopedic and dental implant failure post implantation.

A related complication, arising from the relative motion of bone implants post implantation, is formation of particulates, that is, small particles of synthetic materials (metals, polymers, etc.), whose type and concentration may cause inflammation and toxicity, affecting surrounding cells and tissues locally.

Infection is anathema for all implants and should be prevented at all costs. Biomedical engineering endeavors aimed against infection have focused on designing and formulating materials to prevent infection on, and in the vicinity of, implants. For this purpose, appropriate medications (e.g., antibiotics against bacteria) are either coated on the surfaces or loaded into implantable biomaterials in order to prevent attachment of pathogenic microorganisms on implants and to impede subsequent formation of pathogen colonies protected by biofilms.

Even when they are not frequent, mechanical mismatch of bone implants, implant fatigue, and fracture, and metal implant breakdown post implantation are regrettable but avoidable developments by using judicious design, formulation, and production of bone implants prior to implantation.

9.5 Etiology of Implant Failure Post Implantation

The etiology of implant failure post implantation is comprised of a long list [1] of reasons including wrong or inadequate material-related properties, design flaws and mistakes during product fabrication, inadequate preclinical animal tests, inadvertent damage or contamination during manufacturing, sterilization, packaging, shipping, and storage of implants before delivery to clients at hospitals, contamination, or device damage (which can occur unintentionally at any point, including during implantation surgery), poor patient management, surgery-related mistakes, and patients (a major, unpredictable factor in the implant-related biomedical engineering endeavor). This list provides unquestionable evidence regarding the many aspects which can result in implant failure and emphasizes not only the seriousness of the related complications but also the many endeavors needed to overcome these problems [1]. Because many adverse developments are possible, all key participants – engineers, scientists, manufacturers, surgeons, health-care providers, patient management personnel, and patients – share pertinent concerns and responsibilities.

9.6 Evaluation of Retrieved Failed Implants. Responsibilities of Biomedical Engineers

The comments made so far in this chapter summarize some important aspects of the lessons learned from implant failure post implantation. Such information is collected after failed implants are excised (via another surgical procedure), retrieved, and evaluated. Pertinent analyses use the latest methodologies and instruments available and focus on elucidating the cause(s) of implant failure post implantation. Examination and evaluation of failed implants retrieved from patients provide invaluable information related to safety and efficacy. Only identification of the causes of implant failure can lead to solutions toward improving the design, formulation, and production of the next generation of successful implant biomaterials and medical devices.

It should be noted that implant failure post implantation is not associated with the biocompatibility evaluation of implants prior to implantation (Chapter 8). Implant failures *in vivo* are not planned, are unwanted and unexpected, but they happen. Biomedical engineers and scientists need to (i) elucidate what happened and then correct the aspects which caused implant failures; (ii) need to study in detail failed implants and identify the causes of their failure inside the bodies of animals and humans; (iii) should subsequently apply such information to improve the design, fabrication, and production of biomaterials and medical devices; (iv) establish new criteria for selecting materials for implants; and (v) develop revised protocols and techniques for the various stages of examining failed implants and for evaluating the design, performance, and biocompatibility of implant biomaterials and devices pre- and post-implantation.

9.7 Closing Remarks

In order to succeed, biomedical engineering endeavors aspiring for successful implants post implantation must implement the lessons learned from past failures to identify the parameters which are critical for the performance of implantable biomaterials and medical devices *in vitro* and *in vivo*.

For that purpose, insights emerging from research regarding cell-extracellular matrix interactions, cell signaling processes, cellular bioengineering, tissue wound healing, developmental and systems biology, molecular biology, molecular biochemistry, physiology, pathology, and from all other areas, which are advancing the frontiers of science and engineering, should be integrated into the design, formulation, and production of new biomaterials. It is only through integration of lessons learned from the past with creative new ways of thinking and application to current and future endeavors, biomedical engineers will succeed in producing the best, if not the ideal, implantable biomaterials and medical devices of the future.

Reference

1 Harman MK. 3.1.9 – Medical Device Failure–Implant Retrieval, Evaluation, and Failure Analysis. Editor(s): Wagner WR, Sakiyama-Elbert SE, Zhang G, Yaszemski MJ. *Biomaterials Science (Fourth Edition)*, Academic Press, 2020, Pages 1485–1495. ISBN 9780128161371. https://doi.org/10.1016/B978-0-12-816137-1.00096-9.

10

Hydrogels: Promising, Versatile Biomaterials for Implants

10.1 Introduction

Due to their versatility and promise, hydrogels continue to be utilized for various biomedical applications such as drug delivery, implants, and tissue engineering. Both theoretical/computational and experimental approaches have been used to study hydrogels. The results of past theoretical/computational studies have provided valuable insight regarding the structure and function of this group of biomaterials. Experimental studies confirmed theoretical predictions, which added further information to the importance of hydrogel applications in the biomedical field.

10.1.1 Hydrogel Fundamentals

Hydrogels are three-dimensional, hydrophilic networks (cross-linked polymers), fabricated from homopolymers, copolymers, or macromers [1]. These hydrophilic polymer networks have a large affinity for water but are insoluble due to the formation of either chemical or physical cross-links, which provide structure and physical integrity throughout the network [2, 3]. Both synthetic and natural polymers have been used to synthesize hydrogels and include polyacrylates (such as poly(2-hydroxyethyl methacrylate)), poly(ethylene glycol), poly(vinyl alcohol), poly(lactic acid), poly(glycolic acid), alginate, chitosan, fibrin, collagen, etc. [4]. Hydrogels made from these materials are attractive for many biomedical applications due to their high water content and biocompatibility. In addition, the swelling properties of hydrogels make these materials excellent candidates for various drug-delivery applications.

Concepts of Tissue-Biomaterial Interactions: Fundamentals and New Directions,
First Edition. Rena Bizios and Marissa E. Wechsler.

Companion Website: www.wiley.com/go/Concepts_of_Tissue-Biomaterial_Interaction_1e

10.1.2 Hydrogel Structure and Network Characterization

Hydrogels can be neutral, cationic, anionic, or ampholytic as determined by pendant groups incorporated into the polymer backbone. Ionic gels, whether cationic, anionic, or amphoteric, are generally charged to some extent in aqueous solvents. Hydrogels which deprotonate in the absorbed solvent are anionic, while hydrogels that protonate are cationic. As a result of various pendant group incorporation into the hydrogel backbone, these networks can demonstrate environmentally responsive swelling behaviors. Factors which can affect hydrogel swelling include pH, temperature, and ionic strength.

Several quantitative parameters describe the network structure of hydrogels. The most important parameters used for hydrogel characterization include the polymer volume fraction in the swollen state ($\upsilon_{2,\,s}$), the average molecular weight between cross-links ($\overline{M_c}$), and the distance between cross-links, or mesh size (ξ). These related parameters can be determined either theoretically or experimentally; the two most prominent methods used to obtain these parameters which describe the hydrogel network structure are derived from equilibrium swelling theory and rubber elasticity theory [5].

10.1.3 Equilibrium Swelling Theory

The Flory–Rehner theory [6], used for the analysis of neutral hydrogels, describes swelling by stating that cross-linked polymer gels immersed in a liquid environment reaches equilibrium as a subject of two opposing forces: (i) the thermodynamic force via mixing and (ii) the retractive force of the polymer chains. Analytically, this outcome is described by the Gibbs free energy, where the ΔG_{total}, shown in Eq. (10.1), represents the sum of the contribution of elastic, retractive forces inside the hydrogel by the $\Delta G_{elastic}$; and ΔG_{mixing} as a measure of the compatibility of the polymer and the solvent.

$$\Delta G_{\text{total}} = \Delta G_{\text{elastic}} + \Delta G_{\text{mixing}} \tag{10.1}$$

Differentiation of Eq. (10.1), with respect to the moles of solvent molecules, while maintaining constant pressure and temperature, results in Eq. (10.2), where μ_1 and $\mu_{1,\,o}$ are the chemical potential of the solvent in the hydrogel and pure solvent, respectively. At equilibrium, the difference between the two chemical potentials (solvent outside and inside the hydrogel) must be equal to zero; therefore, differences in the chemical potential due to elastic forces and mixing must balance.

$$\mu_1 - \mu_{1,o} = \Delta\mu_{\text{elastic}} + \Delta\mu_{\text{mixing}} \tag{10.2}$$

The theory of rubber elasticity can be used to determine the change in chemical potential due to elastic forces and, moreover, lead to an expression (Eq. (10.3)) for the average molecular weight between cross-links, $\overline{M_c}$, if the neutral hydrogel is prepared in the absence of solvent. In Eq. 10.3, M_n the number average molecular weight of an uncrosslinked polymer and χ_1 represents the Flory-Huggins polymer-solvent interaction parameter.

$$\frac{1}{\overline{M_c}} = \frac{2}{\overline{M_n}} - \frac{\left(\frac{\bar{v}}{V_1}\right)\left[\ln\left(1 - v_{2,s}\right) + v_{2,s} + \chi_1 v_{2,s}^2\right]}{\left[v_{2,s}^{1/3} - \frac{v_{2,s}}{2}\right]} \tag{10.3}$$

This theory was further modified by Peppas and Merrill [7] for hydrogels prepared in the presence of water (Eq. (10.4)). The change in the chemical potential due to elastic forces now accounts for the volume fraction density of the chains during cross-linking where υ represents the total number of polymer chains, V_1 represents the molar volume of the solvent, and v_2, s represents the polymer volume fraction in the swollen state in water.

$$\frac{1}{\overline{M_c}} = \frac{2}{\overline{M_n}} - \frac{\left(\frac{\bar{v}}{V_1}\right)\left[\ln\left(1 - v_{2,s}\right) + v_{2,s} + \chi_1 v_{2,s}^2\right]}{v_{2,r}\left[\frac{v_{2,s}^{1/3}}{v_{2,r}} - \frac{v_{2,s}}{2v_{2,r}}\right]} \tag{10.4}$$

In Eq. (10.4), $v_{2,\,r}$ denotes the polymer volume fraction in the relaxed state (the state of the polymer immediately after cross-linking).

The presence of ionic moieties within hydrogels adds an additional layer of complexity to the swelling analysis by incorporating a term to account for the total change in Gibbs free energy due to the ionic properties of the network, ΔG_{ionic}. This modification is shown in Eq. (10.5).

$$\Delta G_{\text{total}} = \Delta G_{\text{elastic}} + \Delta G_{\text{mixing}} + \Delta G_{\text{ionic}} \tag{10.5}$$

Upon differentiating Eq. (10.5) with respect to the moles of solvent molecules, while maintaining constant pressure and temperature, Eq. (10.6) can be obtained, where $\Delta\mu_{\text{ionic}}$ represents the change in chemical potential due to the ionic nature of the hydrogel.

$$\mu_1 - \mu_{1,o} = \Delta\mu_{\text{elastic}} + \Delta\mu_{\text{mixing}} + \Delta\mu_{\text{ionic}} \tag{10.6}$$

A more descriptive equation was derived by Brannon-Peppas and Peppas [8] for the swelling characterization of ionic hydrogels. For anionic (Eq. (10.7)) and cationic (Eq. (10.8)) hydrogels, two separate (but equivalent) expressions were obtained. Use of these expressions for analysis of the average molecular weight between cross-links now requires inputs of ionic strength (I) and dissociation constants (K_a and K_b for acids and bases, respectively).

$$\frac{V_1}{4I}\left(\frac{v_{2,s}^2}{\overline{v}^2}\right)\left(\frac{K_a}{10^{-pH}+K_a}\right)^2 = \left[\ln\left(1-v_{2,s}\right)+v_{2,s}+\chi_1 v_{2,s}^2\right] + \left(\frac{V_1}{\overline{v}\overline{M}_c}\right)\left(1-\frac{2\overline{M}_c}{\overline{M}_n}\right)v_{2,r} \times \left[\left(\frac{v_{2,s}}{v_{2,r}}\right)^{1/3}-\left(\frac{v_{2,s}}{2v_{2,r}}\right)\right] \tag{10.7}$$

$$\frac{V_1}{4I}\left(\frac{v_{2,s}^2}{\overline{v}^2}\right)\left(\frac{K_b}{10^{pH-14}+K_a}\right)^2 = \left[\ln\left(1-v_{2,s}\right)+v_{2,s}+\chi_1 v_{2,s}^2\right] + \left(\frac{V_1}{\overline{v}\overline{M}_c}\right)\left(1-\frac{2\overline{M}_c}{\overline{M}_n}\right)v_{2,r} \times \left[\left(\frac{v_{2,s}}{v_{2,r}}\right)^{1/3}-\left(\frac{v_{2,s}}{2v_{2,r}}\right)\right] \tag{10.8}$$

The primary mechanism for solute transport in hydrogels is diffusion through spaces between polymer chains, often referred to as "pores" or "mesh." Depending on the size of the pores, hydrogels can be classified as either nonporous or microporous. In addition, the type and structure of cross-linking throughout the hydrogel can greatly affect its properties (such as porosity) as a result of entanglements, hydrogen bonding, ionic bonding, crystallites, and covalent bonding. The porosity of hydrogels is defined by the structural parameter, mesh size (ξ). The mesh size, or correlation length, determines the average linear distance between cross-links (shown in Eq. (10.9)).

$$\xi = \alpha\left(\overline{r_o^2}\right)^{1/2} \tag{10.9}$$

In Eq. (10.9), the unperturbed end-to-end distance between neighboring cross-links is $\left(\overline{r_o^2}\right)^{1/2}$ (a known value), and the extension ratio (α) can be determined using Eq. (10.10) from the polymer volume fraction in the swollen state ($v_{2,s}$).

$$\alpha = v_{2,s}^{1/3} \tag{10.10}$$

Mesh size can be predicted without the unperturbed distance using Eq. (10.11) if the following values are known: bond length of the polymer backbone (l), the characteristic ratio (C_n), the average molecular weight between cross-links ($\overline{M}_c$), the polymer volume fraction in the swollen state

($\upsilon_{2,s}$), and the molecular weight of repeating units (M_r).

$$\xi = \upsilon_{2,s}^{-1/3}\left(\frac{2C_n\overline{M_c}}{M_r}\right)^{1/2} l \tag{10.11}$$

The ability to tailor the molecular structure of hydrogels using several key parameters, namely the polymer volume fraction in the swollen state ($\upsilon_{2,s}$), the average molecular weight between cross-links ($\overline{M_c}$), and the mesh size (ξ), enables tuning of their diffusive, responsive, and mechanical properties [2].

10.1.4 Solute Diffusion from Hydrogels

Understanding solute transport is the single most important design parameter for hydrogels used as drug carriers. Since solute transport within hydrogels is primarily based on diffusion, Fick's law of diffusion in the x direction can be used to describe this phenomenon (shown in Eqs. (10.12) and (10.13)).

$$j_i = -D_{i,p}\frac{dc_i}{dx} \tag{10.12}$$

$$\frac{\partial c_i}{\partial t} = D_{i,p}\frac{\partial^2 c_i}{\partial x^2} \tag{10.13}$$

Here, c_i represents the concentration, j_i indicates the mass flux of solute i, and $D_{i,p}$ is the diffusion coefficient of solute i in the polymer matrix, p.

It is important to note that several assumptions have been incorporated into Eqs. (10.12) and (10.13). First, these equations describe solute release from a carrier of a thin, planar geometry; equivalent equations for solute release from carries with various other geometries have been derived [9]. In addition, in the form of Fick's law shown in Eqs. (10.12) and (10.13), the diffusion coefficient is assumed to be independent of solute concentration, c_i. Thus, to improve the predictive power of Fickian diffusion, a concentration-dependent diffusion coefficient is used and can be written as Eq. (10.14).

$$\frac{\partial c_i}{\partial t} = \frac{\partial}{\partial z}\left(D_{i,p}\left(c_i\right)\frac{\partial c_i}{\partial x}\right) \tag{10.14}$$

In Eq. (10.14), $D_{i,p}(c_i)$ is the concentration-dependent diffusion coefficient, whose form is dependent on characteristics of the polymer matrix [10].

In typical solute eluting systems, diffusion commonly dictates distribution of solutes throughout hydrogels and delivery to their surrounding environment. Drug release from these swelling-controlled systems (where solute release is mediated by the inward flux of solvent molecules and subsequent swelling of the polymer matrix) can be predicted from a power law relation derived by Korsmeyer and Peppas [11] shown in Eq. (10.15).

$$\frac{M_t}{M_\infty} = kt^n \tag{10.15}$$

Here, M_t and M_∞ are the cumulative amounts of drug released at time t and at equilibrium, respectively; k is a constant dictated by the structural and geometric information regarding the polymer matrix; and n is indicative of the drug release mechanism. Solute release by Fickian diffusion using Eq. (10.15) is approximated using $n = 0.5$.

10.1.5 Hydrogel Types and Applications

Understanding theoretical aspects pertinent to hydrogels has allowed material engineers and scientists to further modulate these systems for various biomedical applications such as drug delivery and tissue engineering. Hydrogels can be synthesized using different techniques resulting in networks of varying geometries (i.e., nanoscale, microscale, bulk gels, etc.), each one appropriate for various biomedical applications.

10.1.5.1 Microscale Hydrogels

Microscale hydrogels are excellent candidates for use in drug delivery, especially controlled release. Materials used as drug-delivery vehicles are to protect therapeutic molecules from undesired degradation, increase the efficacy of drug delivery, and decrease potential side effects. For controlled release, the carriers help maintain the concentration of the contained drug within the therapeutic window over an extended time period, enabling extended time periods to the next delivered drug dose [12, 13]. Controlled release systems are preferred, compared to burst-release, drug-delivery platforms because there is no sharp increase in the delivered drug concentration shortly following administration; such an outcome prevents a potentially toxic response. In addition, the need for less frequent drug administration could potentially promote patient compliance.

Physical and chemical parameters of microgels can be altered to tune their drug-delivery properties. The structure and density of the cross-linking agent used in microgel formulation affect the mesh size, ξ, and thus the rate of the diffusion of water-soluble therapeutic drugs out of the microgel [14, 15]. The overall diameter of the hydrogels also affects controlled drug release, as larger diameters of microgels lead to longer release when the microgels are completely loaded with the therapeutic molecule of interest. Various tissues of interest as delivery sites can also be targeted by altering the overall diameter of hydrogel microparticles. For instance, microgels 1–5 μm in diameter would be best suited to passively target antigen-presenting cells [16] while microgels 10–20 μm in diameter would be better suited to target capillary beds, especially in tumors [17].

10.1.5.2 Nanoscale Hydrogels

Hydrogel nanoparticles in the 1–1000 nm range are highly versatile and tunable systems. In contrast to many other materials, polymers have many unique characteristics enabling them to be used for various drug-delivery applications. Due to their polymeric nature, these materials provide opportunities to control and modulate particle stability, loading efficiencies, release kinetics, and surface modification capabilities. Polymeric nanoparticles have been synthesized from either preformed polymers or monomer(s) using various polymerization techniques, as well as many other methods such as ionic gelation, solvent diffusion, nanoprecipitation, and spray drying [18]. In addition, such nanoparticles can be prepared from both natural and synthetic polymers. In regard to nanoparticle clearance, nanoparticles with a diameters less than 10 nm are rapidly eliminated by the kidneys; in contrast, nanoparticles larger than 200 nm risk activating the complement (and thus the immune) systems [19].

10.1.5.3 Stimuli-Responsive Hydrogels

Stimuli-responsive polymers change conformation or degrade in the presence of a specific condition such as pH, temperature, or enzyme activity. These effects allow the payload to be protected until it is released under the appropriate conditions to which the polymer responds to. Stimuli-responsive materials are used in oral drug-delivery platforms due to their ability to protect the therapeutic molecule in the harsh gastric conditions before releasing it in the small intestine for absorption into the bloodstream [20–24].

In addition to oral drug-delivery formulations, an enzyme-responsive microgel system was studied for other biomedical applications such as pulmonary drug delivery. In one case [25] this platform utilized poly(ethylene glycol) diacrylate microparticles containing specific peptides that degrade in the presence of metalloproteinases, which are overexpressed in certain pulmonary diseases such as lung cancer, tuberculosis, and chronic obstructive pulmonary disease (COPD) [26–28]. Hydrogels are especially useful for pulmonary drug delivery because their size can be precisely engineered for bronchial delivery; moreover, their swelling behavior prevents the microgels from uptake and clearance by macrophages in the alveoli.

10.1.5.4 Responsive Hydrogels in Tissue Engineering

The properties of responsive hydrogels make these materials promising candidates for tissue engineering applications. Current tissue-engineering strategies utilize three main approaches: (i) injection of cells directly into the target tissue of interest, (ii) implantation of cell-containing, three-dimensional tissue structures, and (iii) implantation of scaffold-based delivery systems to supply various drugs and/or proteins to promote cell migration, proliferation, and (if applicable) differentiation [29]. Biomaterial aspects pertinent to these strategies involve several requirements for design and use as tissue-engineering scaffolds; these scaffolds must be biocompatible, biodegradable (in synchrony with new tissue formation), offer similar mechanical properties to that of the targeted tissue, and provide a porous architecture to promote cell penetration/distribution throughout the scaffold and diffusion of nutrients [30]. In addition, these scaffolds must mimic the host microenvironment where cells interact and respond to the biochemical cues and biophysical stimuli pertinent to the surrounding environment [31]. Thus, the material properties of tissue-engineered constructs are critical in modulating subsequent cell functions pertinent to new tissue formation [31]. Challenges of existing tissue-engineering scaffolds include implementing fine-tuned material porosity while maintaining specific mechanical properties, retaining and delivering specific growth factors with ideal release profiles, controlling cell distribution and arrangement within the tissue-engineered construct, and degrading in synchrony with the progress of new tissue formation.

10.1.6 Development of Three-Dimensional Scaffolds with Fine-Tuned Chemistry, Porosity, and Mechanical Properties

The porosity of engineered scaffolds plays a vital role in modulating new tissue formation and function. The porosity of a scaffold material is necessary for providing an environment which allows cell penetration/distribution throughout the scaffold and diffusion of oxygen/CO_2, nutrients and metabolic waste; such aspects are of particular importance in the absence of vascular networks, predominant conditions of scaffolds for various periods of time post implantation. The degree of porosity has an indirect effect on the mechanical properties of the scaffold material, specifically because increase in porosity results in decrease in the mechanical material properties (e.g., stiffness) of the scaffold. In addition, the biological liquid flux can affect many mechanical characteristics of the material, depending on the targeted implantation anatomical location. Furthermore, pore size and pore interconnectivity greatly affect cell functions such as cell proliferation and extracellular matrix synthesis and formation [30].

The ability to modulate hydrogel mechanical/material properties (such as stiffness and pore size) has allowed these materials to be used as tissue-engineering constructs due to their biocompatibility and similarity to the natural extracellular matrix of biological/physiological tissues [32]. Control of specific three-dimensional material properties (such as porosity), however, remains a challenge. In order to obtain an interconnecting, highly porous, three-dimensional scaffold, several fabrication techniques have been employed. Conventional fabrication techniques to control the overall porosity of hydrogels include solvent casting/particle leaching, gas foaming, electrospinning, and freeze drying [32]. However, these methods do not allow for either precise control of the internal, porous architecture within hydrogels or for the formation of complex structures. In addition, the aforementioned conventional methods produce scaffolds with low mechanical/material properties and often require use of toxic solvents. Advanced techniques have been developed to provide control over the microarchitecture and pore features of hydrogels [30]. This outcome has been accomplished through use of various micropatterning and micromoulding techniques such as formation of microchannels. Overall, control of complex features within hydrogels is necessary to guide the development and integration of tissue-engineered constructs with the surrounding biological/physiological environment of their intended application.

10.1.7 Growth Factor Retention, Sequestration, and Delivery

Growth factor signaling is vital to instructing specific cell functions. Cell migration, proliferation, and differentiation can be regulated by cell–cell signaling in response to various growth factors. The critical role of growth factors in maintaining and controlling many basic cell functions underscores their use in tissue-engineering scaffolds to promote and facilitate tissue regeneration. In addition, many growth factors induce cell functions pertinent to therapeutic results such as bone regeneration and neovascularization [29].

Methods of incorporating bioactive growth factors into polymer matrices, such as hydrogels, include chemical immobilization and physical encapsulation. In preparation of polymer-containing growth factor systems, chemical modifications to the polymer matrix and/or physical encapsulation of growth factors may enhance their therapeutic efficiency. In addition, use of polymer matrices as growth factor carriers enables the ability to control specific physical properties of hydrogels to regulate growth factor release rates. Through these mechanisms, optimization of growth factor concentrations and release kinetics can be obtained and utilized to promote tissue regeneration outcomes.

Chemical immobilization of growth factors into biomaterials (such as hydrogels) has been developed using various techniques. The two widely used strategies of conjugating growth factors to hydrogels include physical adsorption and covalent immobilization [29]. Adsorption of growth factors to polymeric matrices is achieved by either secondary or intermediate interactions through intermediate proteins or molecules. Release kinetics of the pre-adsorbed growth factors depends on the association between these bioactive chemicals and the polymer, as well as on environmental conditions such as temperature and pH. Prolonged release of growth factors can be achieved by covalent incorporation of growth factor molecules into the polymer carrier. Bioactive factors can be incorporated into polymers either during the polymerization process or by post chemical or physical modifications. Chemical conjugation of growth factors to the polymer matrix enables formation of strong interactions and provides the potential for localized delivery.

Physical encapsulation of growth factors can be accomplished through various fabrication techniques such as solvent casting and particulate leaching, freeze drying, phase separation, phase emulsion, gas foaming, and *in situ* polymerization [29], with subsequent diffusion-controlled, growth factor release. Further regulation of encapsulation strategies can be achieved by combining fabrication methods (for example, combining burst and sustained growth factor release) and by altering various physical and chemical parameters of the polymer.

References

1 Slaughter B V, Khurshid SS, Fisher OZ, Khademhosseini A, Peppas NA. Hydrogels in regenerative medicine. *Adv. Mater.* [Internet]. 2009 [cited 2016 Jun 21];21(32–33):3307–29. Available from: http://www.ncbi.nlm.nih.gov/pubmed/20882499.

2 Peppas NA, Hilt JZ, Khademhosseini A, Langer R. Hydrogels in biology and medicine: from molecular principles to bionanotechnology. *Adv. Mater.* [Internet]. 2006 [cited 2018 Nov 18];18(11):1345–60. Available from: http://doi.wiley.com/10.1002/adma.200501612.

3 Peppas NA, Bures P, Leobandung W, Ichikawa H. Hydrogels in pharmaceutical formulations. *Eur. J. Pharm. Biopharm.* [Internet]. 2000 [cited 2018 Nov 18];50(1):27–46. Available from: https://www.sciencedirect.com/science/article/pii/S0939641100000904.

4 Neves MI, Wechsler ME, Gomes ME, Reis RL, Granja PL, Peppas NA. Molecularly imprinted intelligent scaffolds for tissue engineering applications. *Tissue Eng. Part B Rev.* [Internet]. 2017 [cited 2018 Nov 20];23(1):27–43. Available from: http://www.ncbi.nlm.nih.gov/pubmed/27484808.

5 Lowman AM, Dziubla TD, Bures P, Peppas NA. Structural and dynamic response of neutral and intelligent networks in biomedical environments. *Adv. Chem. Eng.* [Internet]. 2004 [cited 2018 Nov 20];29:75–130. Available from: https://www.sciencedirect.com/science/article/pii/S0065237703290049.

6 Flory PJ, Rehner J. Statistical mechanics of cross-linked polymer networks I. Rubberlike elasticity. *J. Chem. Phys.* [Internet]. 1943 [cited 2018 Nov 20];11(11):512–20. Available from: http://aip.scitation.org/doi/10.1063/1.1723791.

7 Peppas NA, Merrill EW. Crosslinked poly(vinyl alcohol) hydrogels as swollen elastic networks. *J. Appl. Polym. Sci.* [Internet]. 1977 [cited 2018 Nov 21];21(7):1763–70. Available from: http://doi.wiley.com/10.1002/app.1977.070210704.

8 Brannon-Peppas L, Peppas NA. Dynamic and equilibrium swelling behaviour of pH-sensitive hydrogels containing 2-hydroxyethyl methacrylate. *Biomaterials* [Internet]. 1990 [cited 2018 Nov 21];11(9):635–44. Available from: http://www.ncbi.nlm.nih.gov/pubmed/2090297.

9 Crank J. *The Mathematics of Diffusion.* 2nd ed.Oxford University Press; 1975 [cited 2018 Dec 2]. 414 p. Available from: https://catalog.lib.utexas.edu/record=b2045870.

10 Peppas NA, Narasimhan B. Mathematical models in drug delivery: how modeling has shaped the way we design new drug delivery systems. *J. Control. Release* [Internet]. 2014 [cited 2018 Dec 2];190:75–81. Available from: http://www.ncbi.nlm.nih.gov/pubmed/24998939.

11 Korsmeyer RW, Gurny R, Doelker E, Buri P, Peppas NA. Mechanisms of solute release from porous hydrophilic polymers. *Int. J. Pharm.* [Internet]. 1983 [cited 2018 Dec 2];15(1):25–35. Available from: https://www.sciencedirect.com/science/article/pii/0378517383900649.

12 Liechty WB, Kryscio DR, Slaughter B V., Peppas NA. Polymers for drug delivery systems. *Annu. Rev. Chem. Biomol. Eng.* [Internet]. 2010 [cited

2024 Apr 1];1:149–73. Available from: https://pubmed.ncbi.nlm.nih.gov/22432577/.

13 Vermonden T, Censi R, Hennink WE. Hydrogels for protein delivery. *Chem. Rev.* [Internet]. 2012 [cited 2024 Apr 1];112(5):2853–88. Available from: https://pubmed.ncbi.nlm.nih.gov/22360637/.

14 Peppas NA, Bures P, Leobandung W, Ichikawa H. Hydrogels in pharmaceutical formulations. *Eur. J. Pharm. Biopharm.* [Internet]. 2000 [cited 2024 Apr 1];50(1):27–46. Available from: https://pubmed.ncbi.nlm.nih.gov/10840191/.

15 Kim SW, Bae YH, Okano T. Hydrogels: swelling, drug loading, and release. *Pharm. Res.* [Internet]. 1992 [cited 2024 Apr 1];9(3):283–90. Available from: https://pubmed.ncbi.nlm.nih.gov/1614957/.

16 Wattendorf U, Coullerez G, Vörös J, Textor M, Merkle HP. Mannose-based molecular patterns on stealth microspheres for receptor-specific targeting of human antigen-presenting cells. *Langmuir* [Internet]. 2008 [cited 2024 Apr 1];24(20):11790–802. Available from: https://pubs.acs.org/doi/full/10.1021/la801085d.

17 Wechsler ME, Stephenson RE, Murphy AC, Oldenkamp HF, Singh A, Peppas NA. Engineered microscale hydrogels for drug delivery, cell therapy, and sequencing. *Biomed. Microdevices* [Internet]. 2019 [cited 2019 Mar 29];21(2):31. Available from: http://link.springer.com/10.1007/s10544-019-0358-0.

18 Fonseca-Santos B, Gremião MPD, Chorilli M. Nanotechnology-based drug delivery systems for the treatment of Alzheimer's disease. *Int. J. Nanomedicine* [Internet]. 2015 [cited 2016 Sep 29];10:4981–5003. Available from: http://www.ncbi.nlm.nih.gov/pubmed/26345528.

19 Mitchell MJ, Billingsley MM, Haley RM, Wechsler ME, Peppas NA, Langer R. Engineering precision nanoparticles for drug delivery. *Nat. Rev. Drug Discov.* [Internet]. 2020 [cited 2021 Aug 29];20(2):101–24. Available from: https://www.nature.com/articles/s41573-020-0090-8.

20 Torres-Lugo M, García M, Record R, Peppas NA. Physicochemical behavior and cytotoxic effects of p(methacrylic acid-g-ethylene glycol) nanospheres for oral delivery of proteins. *J. Control. Release* [Internet]. 2002 [cited 2024 Apr 1];80(1–3):197–205. Available from: https://pubmed.ncbi.nlm.nih.gov/11943398/.

21 Carr DA, Peppas NA. Assessment of poly(methacrylic acid-co-N-vinyl pyrrolidone) as a carrier for the oral delivery of therapeutic proteins using Caco-2 and HT29-MTX cell lines. *J. Biomed. Mater. Res. A* [Internet]. 2010 [cited 2024 Apr 1];92(2):504–12. Available from: https://pubmed.ncbi.nlm.nih.gov/19213059/.

22 Steichen S, O'Connor C, Peppas NA. Development of a P((MAA-co-NVP)-g-EG) hydrogel platform for oral protein delivery: effects of hydrogel composition on environmental response and protein partitioning. *Macromol. Biosci.* [Internet]. 2017 [cited 2024 Apr 1];17(1). Available from: https://pubmed.ncbi.nlm.nih.gov/27689827/.

23 Koetting MC, Guido JF, Gupta M, Zhang A, Peppas NA. pH-responsive and enzymatically-responsive hydrogel microparticles for the oral delivery of therapeutic proteins: effects of protein size, crosslinking density, and hydrogel degradation on protein delivery. *J. Control. Release* [Internet]. 2016 [cited 2024 Apr 1];221:18–25. Available from: https://pubmed.ncbi.nlm.nih.gov/26616761/.

24 Durán-Lobato M, Carrillo-Conde B, Khairandish Y, Peppas NA. Surface-modified P(HEMA-*co*-MAA) nanogel carriers for oral vaccine delivery: design, characterization, and in vitro targeting evaluation. *Biomacromolecules* [Internet]. 2014 [cited 2024 Apr 1];15(7):2725. Available from: https://www.ncbi.nlm.nih.gov/pmc/articles/PMC4504688/.

25 Secret E, Kelly SJ, Crannell KE, Andrew JS. Enzyme-responsive hydrogel microparticles for pulmonary drug delivery. ACS Appl Mater Interfaces [Internet]. 2014 [cited 2024 Apr 1];6(13):10313–21. Available from: https://pubmed.ncbi.nlm.nih.gov/24926532/.

26 Ullah MF, Aatif M. The footprints of cancer development: cancer biomarkers. *Cancer Treat. Rev.* [Internet]. 2009 [cited 2024 Apr 1]; 35(3):193–200. Available from: https://pubmed.ncbi.nlm.nih.gov/19062197/.

27 Chang JC, Wysocki A, Tchou-Wong KM, Moskowitz N, Zhang Y, Rom WN. Effect of Mycobacterium tuberculosis and its components on macrophages and the release of matrix metalloproteinases. *Thorax* [Internet]. 1996 [cited 2024 Apr 1];51(3):306–11. Available from: https://pubmed.ncbi.nlm.nih.gov/8779137/.

28 Demedts IK, Morel-Montero A, Lebecque S, Pacheco Y, Cataldo D, Joos GF, et al. Elevated MMP-12 protein levels in induced sputum from patients with COPD. *Thorax* [Internet]. 2006 [cited 2024 Apr 1];61(3):196–201. Available from: https://pubmed.ncbi.nlm.nih.gov/16308335/.

29 Lee K, Silva EA, Mooney DJ. Growth factor delivery-based tissue engineering: general approaches and a review of recent developments. *J. R. Soc. Interface* [Internet]. 2011;8(55):153–70. Available from: https://pubmed.ncbi.nlm.nih.gov/20719768/.

30 Clegg JR, Wechsler ME, Peppas NA. Vision for functionally decorated and molecularly imprinted polymers in regenerative engineering. *Regener. Eng. Transl. Med.* [Internet]. 2017 [cited 2018

Aug 23];3(3):166–75. Available from: http://link.springer.com/10.1007/s40883-017-0028-9.

31 Loh QL, Choong C. Three-dimensional scaffolds for tissue engineering applications: role of porosity and pore size. *Tissue Eng. Part B Rev.* [Internet]. 2013 [cited 2016 Dec 15];19(6):485–502. Available from: http://online.liebertpub.com/doi/abs/10.1089/ten.teb.2012.0437.

32 Annabi N, Nichol JW, Zhong X, Ji C, Koshy S, Khademhosseini A, et al. Controlling the porosity and microarchitecture of hydrogels for tissue engineering. *Tissue Eng. Part B Rev.* [Internet]. 2010 [cited 2016 Dec 16];16(4):371–83. Available from: http://www.ncbi.nlm.nih.gov/pubmed/20121414.

11

Biocompatibility: Past, Present, and Future

Lessons learned from Physiology, Pathology and Biology (Chapters 2, 3, and 4, respectively) provided updated and invaluable information that biomedical biomaterial engineers and scientists need in order to design, formulate, and produce the improved implantable biomaterials and medical devices of the future. In this respect, the primary responsibility of biomedical engineers and scientists is to deliver *biocompatible* products.

Biocompatibility, an integral part (and, thus, of paramount importance) in the field of implantable biomaterials and medical devices, is a fundamental part of the training of students majoring in biomaterials, and it is also relevant to all biomedical engineers and scientists, who are interested, and specialize, in the aforementioned field. The importance of this aspect justifies a short historical review emphasizing several pertinent, current trends and developments, which are briefly discussed in this chapter.

Early definitions of biocompatibility reflected the predominant and pertinent (but in retrospect limited) knowledge of those times. They addressed exclusively biomaterial properties of implantable medical devices, reflected pertinent regulatory agency requirements, and were based on what was then known regarding important physiological processes (such as tissue wound healing) [1]. Implicit was the fundamental requirement that implants should not induce any harm to patients post implantation. Further attention to that requirement resulted in nontoxic, non-pyrogenic, nonallergenic, non-immunogenic, etc., formulations, which improved, but did not resolve, all implantable biomaterial-related complications and failures post implantation.

Since then, use of biomaterials has expanded to now include many other applications, for example, tissue engineering, cell therapy, imaging methods, diagnostic systems, drug delivery, gene delivery, etc. Concurrently, advances in new scientific frontiers (for example, in cellular-, molecular-,

Concepts of Tissue-Biomaterial Interactions: Fundamentals and New Directions,
First Edition. Rena Bizios and Marissa E. Wechsler.

Companion Website: www.wiley.com/go/Concepts_of_Tissue-Biomaterial_Interaction_1e

gene-biology, biochemistry, and cellular engineering) have provided remarkable, new information which is missing from the biocompatibility definitions of the past. For these reasons, the definition of biocompatibility must be modified accordingly and regularly in order to incorporate new trends and developments.

Specifically, the biocompatibility definition must address cellular-, molecular-, and gene-level events and reflect the *in vivo* variability and complexity of the physiological/biological milieux. In addition to "doing no harm", the biocompatible implants are now required to achieve integration with the surrounding cell and tissues by promoting successful, synergistic interactions with, pertinent to, new tissue formation. For this purpose, new criteria incorporating recent advances and knowledge in various areas of biomedical sciences and engineering should be established; moreover, updated methodologies should be developed and used in evaluating the performance of biomaterials, which satisfy the latest scientific requirements [2, 3].

The most important aspect, which should always be kept in mind, is that *biocompatibility of implantable biomaterials and medical devices is a continuously evolving concept*, which must be revised, modified, and refined as and when needed. Such revisions and updates are necessary in order to make biocompatibility relevant to the advances of science and engineering. In the future, biomaterials for implants should be developed based on biocompatibility principles, as well as on replication of the biological/physiological environment of the target cells, tissues, and organs. Biomedical engineers and scientists must keep up with the latest developments in all relevant scientific fields and incorporate the lessons learned in order to develop advanced products needed to help patients now and in the future.

References

1 Williams DF. On the mechanisms of biocompatibility. *Biomaterials* 2008;29(20):2941–53.

2 Peppas NA, Clegg JR. The challenge to improve the response of biomaterials to the physiological environment. *Regener. Biomater.* [Internet]. 2016 [cited 2018 Aug 28];3(2):67–71. Available from: http://www.ncbi.nlm.nih.gov/pubmed/27047671.

3 Anderson JM. Future challenges in the in vitro and in vivo evaluation of biomaterial biocompatibility. *Regener. Biomater.* [Internet]. 2016 [cited 2023 Aug 14];3(2):73–7. Available from: https://dx.doi.org/10.1093/rb/rbw001.

Index

Note: *Italic* page numbers refer to *figures* and **Bold** page numbers reference to **tables**.

Concepts of Tissue-Biomaterial Interactions: Fundamentals and New Directions,
First Edition. Rena Bizios and Marissa E. Wechsler.

Companion Website: www.wiley.com/go/Concepts_of_Tissue-Biomaterial_Interaction_1e

c

r

s

Printed and bound by CPI Group (UK) Ltd, Croydon, CR0 4YY
17/09/2026
14964254-0001